Death, Decomposition, and Detector Dogs

From Science to Scene

Death, Decomposition, and Detector Dogs

From Science to Scene

Susan M. Stejskal, LVT, PhD, DABT

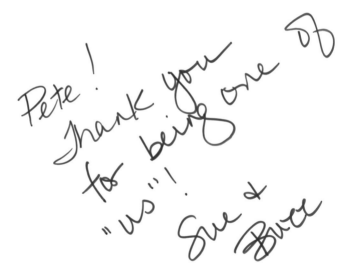

 CRC Press
Taylor & Francis Group
Boca Raton London New York

CRC Press is an imprint of the
Taylor & Francis Group, an **informa** business

CRC Press
Taylor & Francis Group
6000 Broken Sound Parkway NW, Suite 300
Boca Raton, FL 33487-2742

© 2013 by Susan M. Stejskal
CRC Press is an imprint of Taylor & Francis Group, an Informa business

No claim to original U.S. Government works

Printed in the United States of America on acid-free paper
Version Date: 20120725

International Standard Book Number: 978-1-4398-7837-8 (Paperback)

Library of Congress Cataloging-in-Publication Data

Stejskal, Susan M.
    Death, decomposition, and detection dogs : from science to scene / Susan M. Stejskal.
        p. cm.
    "A CRC title."
    Includes bibliographical references and index.
    ISBN 978-1-4398-7837-8
    1. Forensic taphonomy. 2. Human decomposition. 3. Police dogs. 4. Detection dogs.
    5. Crime scene searches. I. Title.

    RA1063.47.S74 2012
    614'.15--dc23   2012009123

Visit the Taylor & Francis Web site at
http://www.taylorandfrancis.com

and the CRC Press Web site at
http://www.crcpress.com

Contents

Acknowledgments

There are many people to thank who helped make this book possible. First is Madeleine Lakatos Fojtik for her technical assistance, illustrations, cover photograph, and design. She is truly a gifted person who I am lucky to call my friend.

I would like to thank the following:

Andy Rebmann, for inspiring those of us in the field and allowing me to use figures from his book; Dr. Lauryn DeGreeff, for sharing information from her doctoral dissertation; Ryan Hunter, for the donation of Hank and Horace, the pigs featured in Chapter 3; and several agencies that shared photos used in Chapter 7. Those agencies included the Portage Police Department, Kalamazoo Township Police Department, Kalamazoo County Sheriff's Department, and the St. Joseph County Sheriff's Department.

Thanks also are in order for those who helped advise and review this project; they include Sgt. John Blue, Lt. Richard Hetu, Sgt./Det. Don McGehee, and, especially, Lt. Marty Johnson.

Special thanks are extended to Sheriff Brad Balk, Undersheriff Mark Lillywhite, and Captain Jason Bingaman (all of the St. Joseph's Sheriff's Department) who supported my growth into law enforcement and by letting us be part of the department.

Thanks as well to Officer Cathy Viverito (New York Police Department) (NYPD, ret.), Officer Benny Colecchia (NYPD), Officer Mark McCallan (NYPD), Lt. Ric Hetu (Michigan State Police, ret.), and Ellen Ponall—they are the true craftsmen of HRD K9 handling. And, to K9 Pal, K9 Joker, and K9 Blaze who bravely served the citizens of New York City.

Thanks to Manny and Reggie Rosenbaum (pops and mom2) who are the true survivors.

Also, thanks to my mom Lois Stejskal who, more than a decade after her passing, continues to inspire me every day. To my dad, retired firefighter George Stejskal, who served the public and inspired me to follow in his footsteps.

Special thanks to my husband, Andrew Rosenbaum, and my son, Joseph, who not only put up with the dogs and my commitments, but have both helped so much throughout the past 10 years. Thanks to both of them for help with photos and technical support.

And, finally, my appreciation goes to Deputy K9 Chili and Deputy K9 Buzz; without them I would have never traveled down this road.

Introduction

The primary goal of any human remains detection (HRD) canine team is to locate a missing person who died from natural or "unnatural" causes. How one does that is the subject of some textbooks and many articles. This book is not a guide on how to train or handle an HRD dog. There are several good books available that can be consulted for this, the first and probably most well known is *Cadaver Dog Handbook* (Rebmann, Sorg, and Carter, 2000). Another is *Buzzards and Butterflies* (Judah, 2008). There have been other very useful books published since then, some of which are listed in the references at the end of the chapters.

So, what is this book about? Basics of canine olfaction describing how dogs are able to do what they do are covered. The intention of this book is to help police investigators and HRD K9 handlers understand some basics of forensic taphonomy—the science of decomposition and how it is affected by the environment. It also covers what is known about the chemical signature of human remains (HR) odor and how environmental factors can affect odor dispersion. Some resources that can be used when planning a deployment are also included.

The role of the "dog handler" is often the misunderstood and undervalued asset of the detection K9 team, very often by the handlers themselves. By understanding some of the science behind just "the dog," a handler should be able to more effectively deploy the K9, helping increase its potential for success.

This book in itself is a tool to help K9 handlers, detectives, and investigators better understand how to use the HRD K9 most effectively as a locating tool.

It is important to point out that this book is a collection of information from a variety of sources and is not intended to be a traditional scientific textbook. Instead it is a *Readers Digest Condensed* version of the science that will hopefully help handlers and investigators do their job better. If more information is wanted, a list of references is included at the end of each chapter.

It is also important to point out that this book primarily addresses decedents who are located outdoors, and the effects of a body either on the surface or buried.

As a picture is worth a thousand words, it is important to acknowledge the contributions of some of the photos used in this book to help further educate those involved in death investigations. Many of the photos were taken by the author; some of them during a visit many years ago to the University of Tennessee's Anthropologic Research Facility in Knoxville. Some photos were taken by the author's husband and son.

References

Judah, J. C. 2008. *Buzzards and butterflies: Human remains detection dogs.* Brunswick County, NC: Coastal Books.
Rebmann, A., E. David, and M. H. Sorg. 2000. *Cadaver dog handbook: Forensic training and tactics for the recovery of human remains.* Boca Raton, FL: CRC Press.

About the Author

Susan Stejskal, PhD, LVT, DABT, is a board-certified toxicologist, licensed veterinary technician, and special deputy and Human Remains Detection (HRD) dog handler with the St. Joseph County (Michigan) Sheriff's Department. With more that 25 years of educational and professional experience, Stejskal has, for the past 10 years, participated in land and water searches throughout Michigan and the central Midwest. She is founder and executive director of Recover K9, a nonprofit organization that provides support for much of her service and educational work. Stejskal  is a member of a scientific working group on Dog and Orthogonal Detector Guidelines (SWGDOG)—a partnership of local, state, federal, and international agencies that develop best practices for use of detector dogs.

Stejskal first partnered in 2000 with HRD K9 Chili (Champion Lone Pine Ms. Chili Dawg, TD, CGC, RN), a miniature, wire-haired

Figure 0.1 AKC Champion Lone Pine Ms. Chili Dawg, TD, CGC, RN, Recipient of a 2011 Honorable Mention for the Award of Canine Excellence: Search and Recovery, American Kennel Club.

Figure 0.2 K9 Buzz, CGC, RN, USPCA cadaver detection-certified, Type I Advanced Disaster Recovery K9.

dachshund. Chili received an honorable mention in 2011 for the Award of Canine Excellence Search and Recovery from the American Kennel Club. Joining Stejskal in 2009 was K9 Buzz, a 3-year-old chocolate Labrador Retriever whom she trained with the Michigan State Police and then certified in cadaver detection by the United States Police Canine Association and as a Type I Advanced Disaster Recovery Dog.

Stejskal's work in toxicology and pathology and her experience as a dog handler led to the development of practical forensic science training for law enforcement dog handlers, detectives, and crime scene technicians. She provides this training for police agencies throughout the country.

Know the Nose

1

Introduction

As part of a detection team, the human remains detection (HRD) K9 handler should be able to describe how a dog is able to work as a locating tool (Figure 1.1). This chapter is a review of the canine olfactory system, and the basics of the anatomy and physiology of scent perception in the dog.

Perception

Although humans use all five senses, we depend very heavily on our sense of sight (Figure 1.2). When we travel through busy streets, we are confronted with moving traffic, lights blinking off and on, and the faces of people passing by. We read body language and facial expressions to guide us in our interactions with those people. Therefore, we depend heavily on our eyes.

Obviously dogs also use their eyes, but they go through their day reading and reacting to their environment through their sense of smell. Instead

Figure 1.1 The tool!

Figure 1.2 Visual overload for people? Is this what it's like for a dog nose?

of enjoying the beauty of a sunrise, a dog enjoys the smell of a fresh pile of bunny poop—something a human thankfully can't really do. Dogs are designed to use their noses.

Anatomy

The olfactory system is basically the same in humans and dogs, but there are obviously some differences that allow dogs to do what they do. The goal is to get air that contains odor chemicals to the cells that detect it and then send the message to the brain where that input is processed. The basic parts of the olfactory system are the nostrils, nasal turbinates, olfactory sensory cells, olfactory nerves, and the brain. The *Cadaver Dog Handbook* by Rebmann, David, and Sorg (2000) provides a detailed description of the canine olfactory system.

First, breathing is a part of olfaction. Air carrying odor chemicals is inhaled through the nostrils or nares (Figure 1.3). When humans breathe, the air goes straight in and then straight out. A dog's nostrils are unique because they are capable of flaring or moving as the dog exhales. As a dog exhales, the exhaled breath is directed off to the side. Unlike humans with our stubby, straight nostrils, the dog doesn't rebreathe much of the same air it just exhaled. There is a fresh supply of air with every breath they take. The nostrils are divided by a septum and so are two separate bony structures.

Figure 1.3 The air carrying odor chemicals is inhaled through the nostril or nares.

Once air passes into the nostrils, it moves through the nasal turbinates, a highly coiled pathway surrounded by bone. The coiled formation of the turbinates creates a large surface area with an extensive blood supply. That blood supply helps to warm, filter, and moisten the air after it's inhaled.

The turbinates are lined with different kinds of cells that help support the olfactory sensory cells. They include the goblet cells and pseudostratified, ciliated, epithelial cells (Figure 1.4). Let's look closer. The goblet cells produce mucus or phlegm. This mucus coats the top surface of the cells in the turbinates. The epithelial cells have cilia or hair-like projections that stick above the cells into the mucus layer. The job of these two kinds of cells is to filter incoming air by trapping dust or other small particles in the mucus, and then the cilia brush or push the mucus down toward the pharynx (near the back of the mouth) where it can be swallowed, sneezed, or coughed out.

The olfactory sensory cells (OSCs) are very different from the other cells (Figure 1.5). They are actually bipolar nerve cells that have cilia on one end while the other end works as a nerve that sends information to the brain. The OSCs interact with incoming odor chemicals, allowing olfaction to happen. How it works will be covered later.

Regular breathing in the dog allows some of the air to get to the OSCs. When a dog pants or breathes through its mouth, even less air gets to the OSCs. Any detection dog handler can describe the changes that they see in

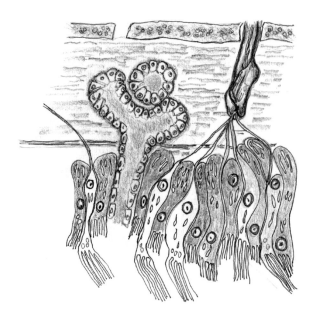

Figure 1.4 Light-colored olfactory sensory cells surrounded by darker support cells and mucous secreting goblet cells. (Adapted from Hole, J. W. 1990. *Human anatomy and physiology*. Dubuque, IA: Wm. C. Brown Publishing, Chap. 12.)

Figure 1.5 Closer view of the sensory cell with cilia at the bottom and nerve at the top leading to olfactory nerves. (Adapted from Hole, J. W. 1990. *Human anatomy and physiology*. Dubuque, IA: Wm. C. Brown Publishing, Chap. 12.)

Figure 1.6 Sniffing *is* different than breathing.

their dog when the dog is onto the odor they are trained to detect; the breathing changes to sniffing (Figure 1.6). Sniffing is described as a "disruption of normal breathing . . . a series of rapid and short inhalations and exhalations" (Correa, 2005). But, compared to regular breathing, a dog's sniffing causes something special to occur. First, the outward puff of air can raise a cloud of dust, freeing the odorant chemicals (Goldblatt, 2010). Sniffing causes the inhaled air to become trapped in the nasal pocket, a cave-like area formed in the nasal turbinates where many of the OSCs are located. A single sniff can cause an accumulation of odor-containing air to become trapped near the OSCs. The combination of new air coming in and collecting in the nasal pocket allows a dog to detect and recognize minute amounts of odor.

Dogs also have a unique nasal airflow pattern that occurs during sniffing; each nostril can get separate odor samples that, after being processed in the brain, allow a dog to localize the source of the odor (Craven, Patterson, and Settles, 2009). This is similar to how we are able to tell where a sound may be coming from with our ears.

Odor Detection and Olfactory Receptor Cells

As previously described, scent or odor is made up of different chemicals trapped in air. The chemicals are volatile, meaning they evaporate easily at room temperature. The chemicals get carried in with air, transfer to the water in the mucus layer in a dog, and then come in contact with the cilia of the OSCs. It is the unique design of the OSCs that lets olfaction take place.

Olfactory sensory cells are actually chemoreceptors—a type of nerve cell that is specially designed to interact with chemicals. These olfactory receptors (OR) are part of the cilia on the OSCs, located for easy access to incoming air.

A simple way to describe how olfaction works is to compare OSCs to taste buds, another type of chemoreceptor (Hole, 1990). In the human, there have been five basic types of taste buds described as being arranged in specific locations on the tongue (Berkowitz, 2011). If salt encounters the "sweet" taste bud, nothing happens. If salt meets a salt taste bud, then the chemical interacts with the receptor and produces an electrical impulse that is sent to the brain. We now detect the taste of salt.

This same explanation can be used with odor detection. A chemical is carried in with inhaled breath, probably as a gas, dissolves and gets picked up in the watery part of the nasal mucus, and comes into contact with the receptor-containing cilia on an OSC. If it is the right receptor for that particular chemical, binding occurs. A "lock and key" process takes place (Figure 1.7). When a chemical in water vapor (the key) gets stuck in the mucus, it reaches the OSC cilia and then the OR (the lock). If the right key fits the lock, the OSC chemistry changes and produces an electrical current or signal. That signal travels up the nerves to the olfactory bulb and then to centers in the brain where the odor is processed.

It appears that not all ORs are the same. Two researchers (Buck and Axel) won a Nobel Prize in 2004 for cloning an olfactory receptor. This work provided a framework to understand how the olfactory system really works. It is believed that each OR is probably unique and has a single chemical receptor. For a chemical to bind to a receptor, it is probably affected by the size, shape, structure, and concentration of the chemical odorant (Lesniak et al., 2008).

Figure 1.7 Key (chemical in air) fits in lock (receptor in olfactory receptor) = open door (or the beginning of the chemical to electrical signaling to brain).

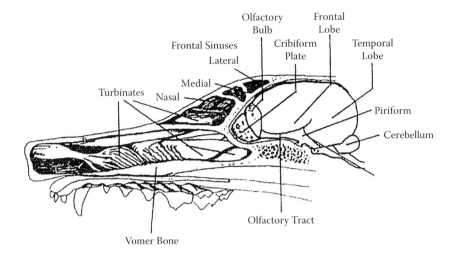

Figure 1.8 Partial dog skull, illustrating nasal cavity and anterior brain. View from side with nasal septum removed to show the turbinates. (From Rebmann, A., E. David, and M. Sorg. 2000. *Cadaver dog handbook: Forensic training and tactics for the recovery of human remains*. Boca Raton, FL: CRC Press. With permission.)

According to Goldblatt, Gazit, and Terkel (2009), each OR "expresses only one receptor protein."

There are >1000 genes that control the ORs of the OSCs in the dog, meaning there is a great degree of variation seen in the OR genetic code among individual dogs and breeds (Tacher et al., 2005). The number and types of receptors vary widely among dogs.

The electrical information is carried from the OSCs by olfactory nerves to the olfactory bulb (OB) in the brain (Figure 1.8). The OB, described as a relay station for electrical signal conduction, has several layers and types of cells involved in conduction and processing of olfactory information (Wei, Zhang, and Guo, 2008). The brain processes and evaluates incoming information and identifies it as a recognizable odor by using basic pattern recognition to interpret the chemical signature (Anon., nd). Button (1990) found that the brain processes what it smells into a scent picture that fades or disperses or fluctuates over time.

The process of olfaction is described in detail in several references (e.g., Helton, 2010; Pearsall and Verbruggen, 1985; Syrotuck, 2000).

Species and Breed Differences

Interspecies Variations: Differences between Dogs and People

There are several differences between the human and dog olfactory systems, which explains the differences in olfactory acuity or the sensitivity of the sense of smell.

The first is the size of the nasal cavity and the amount of air that can be inhaled. According to Pearsall and Verbruggen (1985), German shepherds can breathe in five times more air than a human. Figure 1.9 shows the small nasal capacity in the human. Dogs have more OSCs than humans; estimates of 5 million in the human compared to 125 million in the dachshund, 225 million in the German shepherd, and 300 million in the bloodhound (Coren and Hodgson, 2011).

The human olfactory epithelium containing OSCs covers an area estimated to be about the size of a postage stamp (1.6 in.²) compared to about 26 in.² in the dog (Kaldenbach, 1998). These differences alone provide a simple explanation of why the dog's olfactory system performs better than a human's.

To accommodate the larger number of OSCs, the dog's olfactory bulb is about 40 times larger than in the human (Correa, 2005). This is probably because it takes more brain power to process all of the information that a dog picks up. This system is designed to provide a much higher degree of sensitivity or, as described by Theby (2010), a "higher resolution of the scent." The human's olfactory system is like a camera with a small imaging capacity and a small processing center, while the dog's olfactory system is like the latest and greatest digital camera. In comparison, the human's system would result in photos of low quality and poor focus compared to the dog's with high resolution and detail.

These differences result in a human being able to only smell a mixture of odors compared to the dog that can smell a range of distinct and different

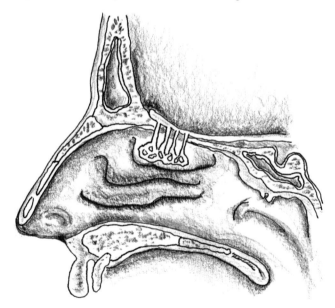

Figure 1.9 Humans and their small nasal capacity. (Adapted from Hole, J. W. 1990. *Human anatomy and physiology.* Dubuque, IA: Wm. C. Brown Publishing, Chap. 12.)

scents. People smell pizza while a dog smells dough, tomato sauce, pepperoni, oregano, and two different types of cheese. And, the dog can tell where the smells are coming from.

Dogs do not suffer from olfactory fatigue like humans do. When we first smell something, it may make quite an impression on us, but as time goes on, our ability to keep smelling or perceiving that smell decreases. This is not the case with the dog.

The dog has something else that a human doesn't: the vomeronasal gland (VNG), also known as the Jacobsen's gland. The VNG is actually a pair of long, fluid-filled sacs that opens into either the mouth or the nasal cavity (Correa, 2005). The glands are located in the area above the roof of the mouth, well behind the upper canines near the vomer bone (Figure 1.8). The sensory cells in the VNG are different than those in the nasal cavity and appear to work differently as well. It appears to be an accessory olfactory organ that allows dogs to identify scent and possibly to identify pheromones, also known as "hormonal chemicals" (Snovak, 2003).

Intraspecies Variation: Differences between Dogs

There is continued debate on whether one breed is better than another (Figure 1.10). Whatever the discussion, dogs have more OSC than humans. A

Figure 1.10 Whatever the breed, dogs are able to smell far better than people.

2007 decision by a court in Texas concluded that the breed of the dog is "not a determinative of a cadaver dog's qualification" (Ensminger and Papet, 2012). Early research showed that there are big differences between the genetic make-up of the OR between breeds. There are about 1300 genes that control what an OR can "look" like and what chemical it may bind with. The OR is actually a strand of molecules that snakes through the membrane of the cilia and has seven loops on the inside and outside of the membrane (Lesniak et al., 2008). These seven loops are called transmembrane domains (because they cross the cilia membranes) and, apparently, each of these domains has a site that has selective binding affinity or binds with specific chemicals. Three of these domains (TM3, 4, and 5) are believed to be involved with odor discrimination.

Tacher et al. (2005) researched how much variation there is in the genes that control OR in 95 unrelated dogs of 20 breeds. They reported a lot of genetic variation. The study showed that some parts of the genetic code (alleles) were breed specific and did not appear significantly in the general dog population. According to their research, the authors found more genetic variation in boxers than in poodles. They concluded that the high degree of genetic variation in ORs could explain why boxers showed less olfactory acuity than poodles.

Another research project compared the genetic makeup of several dogs to their operational success (Lesniak et al., 2008). The researchers found some of the OR genes showed up to seven different places on the receptors that had specific alleles (parts on genes), which "might predispose dogs to detecting defined odors." It was felt that there could be a relationship between certain genotypes and "the ability of more accurate scent detection of particular volatile organic compounds (VOCs)." In this study, they were unable to identify a specific gene that was linked with great scenting ability, but did identify sites on "2 or more genes that may play a role in the effectiveness of odor discrimination."

Researchers also reported that there are differences in the size of the olfactory bulb in dogs of different ages (Wei, Zhang, and Guo, 2008). They reported that the OB in a one-month-old puppy continues to develop through adulthood, and so it is likely that their sense of smell also continues to develop. By exposing a dog to specific odors while training in detection (trailing, tracking, substance detection, etc.), it is likely that their olfactory system develops further, possibly making them eventually better at detecting the odor chemical than other dogs not exposed to those odors (Eckenrode et al., 2006).

Dogs have partnered with man for a long time, allowing the use of their olfactory acuity to help better our lives. We are all familiar with the use of dogs to detect explosives, narcotics, accelerants, and contraband. Scientific studies also have shown that dogs are able to detect a number of different types of human cancers, including prostate (Cornu et al., 2011) and ovarian cancers (Gilden, 2008).

Laboratory testing has been done to measure a dog's olfactory acuity or sensitivity. One study showed that dogs can detect levels of the chemical, N-amyl acetate, down to levels of 1.1 parts per trillion (ppt) (Walker et al., 2006). In a project designed to evaluate canine detection of land mines, researchers found that dogs were capable of detecting at least 1 part per billion (ppb) of 2-4 DNT (an explosive) in air (Waggoner, 2001).

To explain the olfactory acuity or sensitivity in dogs, it is important to be able to describe parts per million, parts per billion, and parts per trillion. The following example may help.

1 part per million (ppm) is the same as 1 second in one year.
1 part per billion (ppb) is the same as 3 seconds in 100 years.
1 part per trillion (ppt) is the same as 3 seconds in 100,000 years.

Orthogonal Devices

The Scientific Working Group on Detection Dogs and Orthogonal Device Guidelines was formed to develop consensus-based guidelines for detector dogs and their handlers (SWGDOG, 2011). The term *orthogonal* refers to devices developed to detect substances that dogs are trained to locate. This generally includes or refers to field-based detectors built to locate substances, such as explosives or chemicals.

One study described an electronic nose developed in 2001 by a university that was capable of detecting at least 500 parts per trillion of explosives in air, but had problems performing consistently in different types of environmental conditions (Waggoner, 2001). Another more recent detector developed is the LABRADOR, short for "light-weight analyzer for buried remains and decomposition odor recognition" (Page, 2010; Larsen, Vass, and Wise, 2011). It is designed to be used in the field at the soil surface to detect chemicals released from decedents in shallow graves, 1.5 feet to 3.5 feet deep. The journal articles stated that, although the "device is not as sensitive as innate canine olfactory capabilities, one advantage of LABRADOR is that it can detect and alert the operator to the *amount* of odor present." Although a dog doesn't indicate the amount of odor will indicate a "yes or no" result, an experienced handler recognizes when the dog first encounters odor, shows interest, and moves toward areas of higher concentration of odor.

Other orthogonal devices will likely be built in the future, but it is unlikely that the HRD dog will be replaced as a locating tool.

Summary

To review, the olfactory system of the dog has a much higher or greater degree of sensitivity than the human for a number of reasons. Dogs have longer and more coiled turbinates with a larger surface area containing many more olfactory sensory cells than humans. Dogs have a more developed and larger olfactory bulb in the brain than the human. Dogs have a capacity to breathe in more air per breath than humans. Dogs are built to not rebreathe as much air as humans, and dogs can differentiate a spectrum of chemicals in each breath. Dogs also are capable of locating the source of odor and can be trained to be reliable and effective locating devices, making the job of law enforcement easier.

Conclusions

At this point in time, it is generally accepted that there is a lot of genetic variation in ORs between breeds and individual dogs. It is recognized as well that a dog's sense of smell is only part of what it takes to be a successful detection dog. The dog's willingness to search in stressful situations, often with many distractions, while partnering with a handler is also a large part of it.

So, there are still many questions:

- Is olfactory sensitivity based on genetics, personality, or training?
- Is it possible that different dogs have different receptors that detect different chemicals?
- Do some breeds have OR keyed for drug detection while others have more ORs for HR detection?
- Does a dog develop more receptors or increase the function of specific receptors with exposure to specific odorants during training?
- Does the brain become more efficient at processing electrical signals coming in from the OR?

The answers are coming slowly but surely. It will be exciting to follow the development of knowledge and see how it applies to better selection, better training, and better handling of detection dogs.

References

Anon. n.d. Olfaction: The sense of smell. Online at: http://users.rcn.com/jkimball.
    ma.ultranet/BiologyPages/O/Olfaction.html (accessed April 26, 2011).

Berkowitz, B. 2011. Taste buds are just one reason why we love some foods and hate others. *The Washington Post* April 25. Online at: http://www.washingtonpost.com/national/science/taste-buds-are-just-one-reason-why-we-love-some-food (accessed May 10, 2011).

Button, L. 1990. *Practical scent dog training*. Loveland, CO: Alpine Publications.

Coren, S., and S. Hodgson. 2001. Understanding a dog's sense of smell. Online at: http://www.dummies.com/how-to/content/understanding-a-dogs-sense-of-smell.navId-323759 (accessed April 9, 2011).

Cornu, J., G. Cancel-Tassin, V. Ondet, C. Girardet, and O. Cussenot. 2011. Olfactory detection of prostate cancer by dogs sniffing urine: A step forward in early diagnosis. *European Urology* 59: 197–201.

Correa, J. E. 2005. The dog's sense of smell. *Alabama Cooperative Extension System—UNP-66.* Online at: www.aces.edu (accessed January 12, 2011).

Craven, B. A., E. G. Paterson, and G. S. Settles. 2009. The fluid dynamics of canine olfaction: Unique nasal airflow patterns as an explanation of macrosmia. *The Royal Society.* Online at: http://rsif.royalsocietypublishing.org/content/7/47/933.abstract (accessed April 26, 2011).

Eckenrode, B. A., S. A. Ramsey, R. A. Stockham, G. J. Van Berkel, K. G. Asano, and D. A. Wolf. 2006. Performance evaluation of the scent transfer unit™ (STU-100) for organic compound collection and release. *Journal of Forensic Science* 51 (4): 780–789.

Ensminger, J. J., and L. E. Papet. 2012. *Police and Military Dogs: Criminal Detection, Forensic Evidence, and Judicial Admissability*. Boca Raton, FL: CRC Press, Chap. 19.

Gilden, J. 2008. Trained dogs able to detect ovarian cancer's specific scent. *Medical News Today* June 27. Online at: http://www.medicalnewstoday.com/printer-friendlynews.php?newsid=113085 (accessed June 15, 2008).

Goldblatt, A., I. Gazit, and J. Terkel. 2009. Olfaction and explosives detector dogs. In *Canine ergonomics: The science of working dogs,* ed. W. S. Helton, pp. 135–174. Boca Raton, FL: CRC Press/Taylor & Francis.

Helton, W. S. 2009. *Canine ergonomics: The science of working dogs.* Boca Raton, FL: CRC Press.

Hole, J. W. 1990. *Human anatomy and physiology*. Dubuque, IA: Wm. C. Brown Publishing, Chap. 12.

Judah, J. C. 2008. *Buzzards and Butterflies: Human remains detection dogs.* Stanger, South Africa: Coastal Books.

Kaldenbach, J. 1998. *K9 scent detection: My favorite judge lives in a kennel.* Calgary, Alberta, Canada: Detselig Enterprises Ltd.

Larson, D. O., A. A. Vass, and M. Wise. 2011. Advanced scientific methods and procedures in the forensic investigation of clandestine graves. *Journal of Contemporary Criminal Justice* XX (X): 1–34.

Lesniak, A., M. Walczak, T. Jezierski, M. Sacharczuk, M. Gawkowski, and K. Jaszczak. 2008. Canine olfactory receptor gene polymorphism and it relation to odor detection performance by sniffer dogs. *Journal of Heredity* 99 (5): 518–527.

Page, D. 2010. LABRADOR: The new alpha dog in human remains detection? *Forensic Magazine.* Online at: http://www.forensicmag.com/article/labrador-new-alpha-dog-human-remains-detection?page=0,0 (accessed June 11, 2011).

Pearsall, M. D., and H. Verbruggen. 1982. *Scent: Training to track, search, and rescue.* Loveland, CO: Alpine Publications.

Rebmann A., E. David, and M. Sorg. 2000. *Cadaver dog handbook: Forensic training and tactics for the recovery of human remains.* Boca Raton, FL: CRC Press.

Snovak, J. E. 2003. *Barron's guide to search and rescue dogs.* Hauppauge, NY: Barron's Educational Series.

SWGDOG (Scientific Working Group on Detector Dogs and Orthogonal Device Guidelines). Online at: www.swgdog.org (accessed April 4, 2003).

Syrotuck, W. G. 2000. *Scent and the scenting dog.* Mechanicsburg, PA: Barkleigh Publishing.

Tacher, S., P. Quignon, M. Rimbault, S. Dreano, C. Andre, and F. Galibert. 2005. Olfactory receptor sequence polymorphism within and between breeds of dogs. *Journal of Heredity* 96 (7): 812–816.

Theby, V. 2010. *Smellorama: Nose games for dogs.* Dorchester, U.K.: Veloce Publishing Ltd.

Waggoner, L. P. 2001. Canine olfactory sensitivity and detection odor signatures for mines/UXO, testing support for Tuft's Medical School E-nose, and fate and effects team participation (MDA972-97-1-00026 Report). Auburn, AL: Auburn University.

Walker, D. B., J. C. Walker, P. J. Cavner, J. L. Taylor, D. H. Pickel, S. B. Hall, and J. C. Suarez. 2006. Naturalistic quantification of canine olfactory sensitivity. *Applied Animal Behavior Science* 97: 241–254.

Wei, Q., H. Zhang, and B. Guo. 2008. Histological structure difference of dog's olfactory bulb between different age and sex. *Zoological Research* 29 (5): 537–545.

Forensic Taphonomy

2

Breaking Down
Is Hard to do

Introduction

The purpose of this chapter is to provide a background for human remains detection (HRD) dog handlers and/or those involved in death investigations. It is hoped that after reading this, one will be able to understand the basics of normal cellular function; of decomposition; products of decomposition; and, finally, to describe taphonomy and its application to investigations.

This information summarized in this chapter comes from a number of different textbooks that can provide detailed information about the process of forensic taphonomy.

Taphonomy Defined

Taphonomy has been defined as the "study of a decaying organism" and comes from Greek words *taphos* and *nomos* meaning burial and a system of laws (*Merriam-Webster Dictionary*, 2011). The practical definition involves what factors affect and influence decomposition of an organism. There are different fields of taphonomy including biological, cultural, and geological (Haglund and Sorg, 1997). The application of forensics to taphonomy involves looking at the specifics of postmortem processes and what may have caused it to proceed the way it did. Much of the information summarized in this chapter is from Haglund and Sorg (1997) and Tibbett and Carter (2008). The basics of how this occurs follows.

The Body

The body is made up of cells and those cells may be organized with similar or like cells into tissues, which make up organs, organs with similar functions are designated as systems, and systems make up the entire body (Figure 2.1).

Figure 2.1 Bodies are made up of ears, noses, different systems, tissues, and cells.

Decomposition takes place in the same way, by affecting cells, systems, and, then finally, the body.

Normal Cell Composition

Every cell in the body has the same basic structure, which includes a nucleus, mitochondria, cell membrane, and other parts that one may remember from high school Biology class (Figure 2.2). The nucleus is the brains of the cell, coordinating what, how, and when things happen. Mitochondria are the power house of the cell, supplying energy to keep things running. The main energy source of the cell is adenosine triphosphate or ATP. There are many other things inside the cell that help it to maintain itself and coordinate the very complex biochemical processes that keep a cell alive. One of the most important structures, however, is the cell membrane.

In the "typical" mammalian cell, the membranes also are complex (Figure 2.3). The membrane is not just a wall holding things inside the cell, instead it is a very complicated structure that is actually made up of "bipolar lipid layers" (Hole, 1990). These can be compared to an exterior house wall—the drywall on one side and the vinyl siding on the other make up the lipid or fat layers, which are somewhat resistant to water. The layer in between is like the house insulation. Through this wall, there are many different types of channels, pores, and receptors that control what passes into and out of a cell, like the windows and doors of a house. The cell "windows" and "doors" open and close by a very complex system that controls the internal and external chemistry of the cell. Normal cell function requires controlled temperature, the right amount of water, and adequate energy to fuel the system.

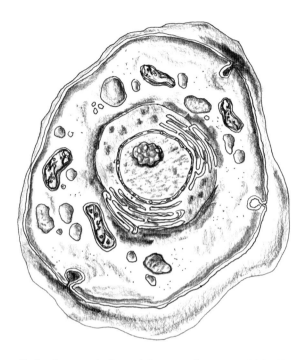

Figure 2.2 Cellular basics. (Adapted from Hole, J. W. 1990. *Human anatomy*. Dubuque, IA: Wm. C. Brown Publishing.)

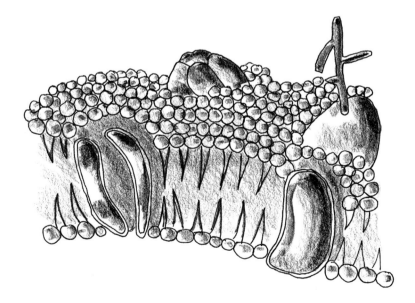

Figure 2.3 The cell membrane is like the wall of a house with doors (pores) and windows (receptors). (Adapted from Hole, J. W. 1990. *Human anatomy*. Dubuque, IA: Wm. C. Brown Publishing.)

The first requirement for normal cell chemistry is a steady, constant temperature. In humans, the constant, internal temperature is about 98.6°F or about 37°C. Many of the biochemical reactions that must take place to keep a cell alive and functioning depend on just the right temperature. Increases or decreases in temperature can speed up or slow down these chemical processes.

The second requirement is water (Figure 2.4). A balance of water inside and outside of a cell (intracellular and extracellular) is essential for the life of the cell. Too much water inside and the cell can become flooded; too little and the cell can dry out. Water also acts as a buffer to chemically moderate cell function, by controlling the acidity or alkalinity inside and outside of a cell.

Note that acidity is defined as the hydrogen-ion concentration in gram equivalents per liter of solution (*Merriam-Webster Dictionary*, 2011) or, more practically, a number on a scale of 0 to 14. The lower the number, the more acidic a solution, and, thus, the higher the number, the solution is less acidic and more basic or alkaline. Table 2.1 shows the range of acidity/alkalinity found in a number of household products.

The third requirement is energy (Figure 2.5). The basic form of energy used by cells is adenosine triphosphate (ATP), which fuels the chemical

Figure 2.4 Cells and bodies require water.

Table 2.1   Range of pH and common things

pH	Thing
1	
2	Battery acid
3	Lemon juice
4	Vinegar
5	
6	Milk
7	
8	Baking soda
9	
10	Milk of magnesia
11	
12	Ammonia
13	Lye
14	

Figure 2.5 Cells require energy as cars require gas. (Photo by J. K. Rosenbaum.)

reactions that open and close the channels, pores, and receptors in the cell membrane. Energy is required to move things in and out of the healthy cell to maintain the intracellular and extracellular balance.

Cellular Decomposition

If conditions change and there is no longer the right temperature, the right amount or balance of water, and/or enough energy to power these processes, cells can die. If cells are too hot or too cold, they can die. If there is an imbalance in cell water (affecting the pH or acidity), the cell can die. If there is not enough energy, cells can die. When cells die, they go through a process of breakdown or decomposition. These changes can begin as early as four minutes after death (DeGreeff and Furton, 2011).

Decomposition is a complex process, and is highly dependent on environmental factors. Not all types of cells die at the same time and in the same way. There is a *huge* difference in decomposition of different types of cells in the human body, all dependent on very specific conditions. Cells that normally contain a large number of proteins that break down other proteins and fats (enzymes called proteases and lipases) will generally decompose faster than other types of cells (Clark, Worrell, and Pleuss, 1997).

At the cellular level, decomposition goes through a series of different processes or stages. The first part of decomposition is reversible; if temperature, water balance, or energy level is returned to normal, the cell may live. If conditions do not return to normal, the cell can go past the point of no return and die. Basically, the following can occur.

Part 1: Reversible:

- For some reason, there is not enough energy (ATP) to fuel the cell and the normal biochemical functions slow down or stop. Not all processes are affected in the same way, so different things may happen at different times in different kinds of cells.
- Because there isn't enough energy to power the chemical processes needed to maintain the balance inside the cell, the acidity levels increase (pH decreases).
- As the receptors, pores, and channels are no longer able to pump waste products out of the cell or bring nutrients in, the cell becomes flooded.
- If conditions don't improve, the cell can progress to Stage 2.

Part 2: Irreversible

- The pressure from flooding inside the cell increases and starts to push and stress the cell membrane (like an overfilled balloon).
- The cell membranes begin to break apart and leak, allowing enzymes that normally break down things inside the cell to leak outside the cell.
- As the enzymes leak, they can begin to break down neighboring cell membranes causing the cells to detach from each other.
- As groups of cells die, the tissues become paler in color and fragile, often slipping apart.

Systemic Decomposition

It may be easier to understand decomposition at the systemic or whole body level. Changes in body temperature affect both the cellular and systemic levels. Changes in circulation or blood supply can affect the flow of nutrients and oxygen into cells and waste products and carbon dioxide (CO_2) from cells. Changes in pH can have a direct effect on the health of cells, organs, and systems.

Stages of Decomposition

Decomposition has been described in many ways. One of the more descriptive includes 10 different stages (Table 2.2). One of the simplest classifications has three stages: immediate (cellular), putrefactive, and postputrefactive. What is important to understand is that not all bodies will go through all the stages in the same way and that the changes that can take place are very dependent on the type of environment to which the body is subjected. The environmental effects will be described in further detail in Chapter 3.

So, to understand taphonomy, we will work through each of the stages. This chapter will get a little deep into decomposition, but it is necessary for one to understand that follows in Chapters 3 and 4.

Table 2.2 Detailed stages of decomposition*

Stage	Description
	Putrid
Stage 1	Early putrid odor, lividity fixed, rigor waning, tissues tacky
Stage 2	Green abdominal color, hemolysis, intense livor, no rigor, early skin slippage, drying of nose, lips, and fingers
Stage 3	Tissue gas, prominent hemolysis, tissues soft and slick, skin slippage
	Bloating
Stage 4	Early body swelling, discoloration of the head, no discoloration of the trunk, gas in heart, marbling, formation of gas bullae
Stage 5	Moderate swelling, discoloration of head and trunk
Stage 6	Maximal body swelling
	Destruction/Decay
Stage 7	Release of gases, exhausted putrified soft tissue, total destruction of blood
Stage 8	Partially skeletonized, mummification, or formation of adipocere
Stage 9	Skeleton with ligaments
Stage 10	Skeleton without ligaments (no soft tissue)

* Adapted from Rebmann, A., E. David, and M. H. Sorg. 2000. *Cadaver dog handbook; Forensic training and tactics for the recovery of human remains.* Boca Raton, FL: CRC Press.

Stage 1: Immediate

Immediate decomposition, described above, is what takes place at the cellular level. As was discussed, there are two stages: reversible and nonreversible. Early changes that can be seen in a body include a pallor to the skin and relaxation of some muscles (Clark, Worrell, and Pleuss, 1997). As time continues, the decomposition progresses.

Stage 2: Early

The most obvious signs of the progression of decomposition are referred to as the "mortises" or algor mortis, livor mortis, and rigor mortis. Each is briefly discussed below.

Algor Mortis

Algor mortis simply refers to the normal cooling of the body that takes place after death. Because the heart is no longer beating and the blood circulating, the body begins to cool from the average 98.6°F. The average rate of coding is estimated to be a decrease of about 1.5° per hour after death. This, of course, depends on what the temperature of the person was at the time of death *and* the ambient temperature in the area of the decedent. If the decedent had a 104°F fever or died outside in a blizzard, the 1.5 degrees/hour estimation would certainly be inaccurate.

The scene from many of the television crime shows—as the thermometer is pulled out from the body, the crime scene tech declares, "Time of death was 2 hours and 37.5 minutes ago"—is, well, television—both wrong and misleading.

Livor Mortis (aka Livor or Lividity)

As the postmortem interval (PMI) increases, the pallor of the skin can progress to what is described as the bluish color of death; the literal meaning of livor mortis. The term refers to the gravitational pooling of blood after death. After the heart stops beating, the blood becomes stagnant and follows the laws of gravity, settles in the lowest parts of the body. Lividity is generally seen first about two hours after death. According to one author, livor can actually start before death in patients with heart disease, likely due to decreased circulation of blood to all parts of the body (Snyder-Sach, 2001). The initial color of livor is generally a pink to red color and then darkens with time. Areas where the blood has been pushed out of the blood vessels due to compression against a solid surface can appear white (Figure 2.6).

The state of livor can be checked by pushing against an area that is pinked or reddened. After pressing on the skin, the area blanches (turns pale or white), it is likely that the body is in the first few hours after death (Snyder-Sachs, 2001). If the body is moved during this time (within a few hours after

Figure 2.6 The deeper colors reflect blood settling and concentrating in lower parts of the body, due to gravity. The decedent had been moved to show signs of livor. (Photo taken at University of Tennessee's Anthropologic Research Facility and courtesy of the author.)

death), the pattern of livor (where the blood has settled) can redistribute. A good observer may be able to tell the difference in livor patterns.

Livor or lividity becomes "fixed" after as early as four to six hours after death (Clark, Worrell, and Pleuss, 1997), but usually by 12 hours after death (Snyder-Sachs, 2001). This occurs when the blood vessels "close down" and the blood can't shift with gravity anymore (Snyder-Sachs, 2001). If a decedent with fixed livor is moved and there is a marked difference between the blanched and reddened area, one could deduce that the decedent had been dead for at least 12 hours and that the body had been moved.

Livor will persist until the blood vessels break down and bacteria moves in and helps the blood to disperse throughout the body.

Rigor Mortis

Rigor is probably one of most recognized postmortem signs and is simply the stiffening of muscles (Figure 2.7). What most people don't realize is that rigor is something that comes and goes. How does this happen? Well, initially, skeletal and smooth muscles relax shortly after death. This is why feces and/or urine are often released at or after death (Clark, Worrell, and Pleuss, 1997).

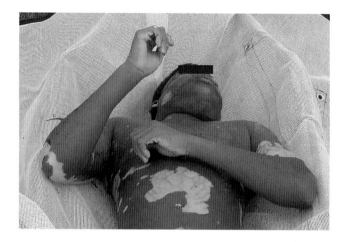

Figure 2.7 Rigor mortis is transient muscle contraction due to lack of energy.

But, what about at the cellular level? Remember ATP? Discussed before, ATP is required to maintain the integrity of cell membranes and for normal cell function. In muscles, ATP is required to contract *and* relax muscle fibers. If there is no ATP, the muscle fibers contract and shorten, becoming stiff. A state of persistent contraction persists.

Some believe that rigor mortis begins with the muscles in the head, continuing down to the chest, and then finally to the limbs. What is more likely is that changes in the smaller muscles (like in the face), are seen first. Rigor is generally seen for the first time about two to six hours postmortem and will involve muscles throughout the entire body over the next six hours.

It is important to understand that persistent muscle contraction does not persist. The "rock solid" state lasts until the chemicals in the muscles are consumed, then degrade, causing the cell membranes to break down and pull apart from each other (Snyder-Sachs, 2001). As this happens, the muscles loosen and relax, usually seen in the opposite order in which rigor sets in. Generally, rigor can persist for somewhere between 24 to 48 hours and then fades away.

The actual duration of rigor depends on a number of factors, mainly temperature. As with algor mortis, the premortem temperature of the deceased and the environmental temperature can affect the development and duration of rigor. If a person had a fever, vigorous activity, or exertion just before death, the onset of rigor may be faster due to the presence of lactic acid in and around the muscle cells. Cold ambient temperature can accelerate and prolong rigor. In hot temperatures, rigor may never occur. The increased temperature can cause an increase in the chemical reactions, making the cells decompose quickly and detach from each other. If this happens, rigor may never set in.

Stage 3: Putrefaction

Putrid—what most people associate with this stage is color, bloat, and odor (Figure 2.8). This is the stage which bacteria really come to the forefront in assisting with breaking down a body. The active microbial population (generally bacteria) at this point is primarily the type that grows when there is low to no level of oxygen present. This is known as an anaerobic condition. These bacteria usually come from either the large intestine or soil or both, and attack cells in the body causing the breakdown of carbohydrates, proteins, and fats. Bacteria will be discussed further in Chapter 4.

Colors of Putrefaction

The easiest way to understand the color changes of decomposition are is to compare it to bruising in live people. In bruising, an injury causes blood vessels to leak, which creates free blood under the skin (hematoma). It may take a while for a bruise to appear, but when it does, it may first look like a black and blue spot. As the extravascular (outside of the vessels) blood breaks down, there are many colors associated with hemoglobin (the oxygen-carrying protein of red blood cells). As the postmortem interval increases, time goes on, the accumulation of carbon dioxide (CO_2) in the decedent's blood causes it to slowly become acidic and to clot throughout the body (Clark, Worrell, and Pleuss, 1997). The hemoglobin (with no oxygen) breaks down to bile pigments causing a progression of colors from reds and purples to browns and blacks (Figure 2.9). A persistent cherry red color may indicate

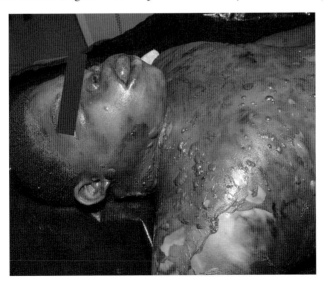

Figure 2.8 Putrefaction: Increase in bacterial populations that produces gas as evidenced by gas bubbles under the skin and the protruding tongue.

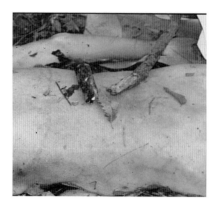

Figure 2.9 Changes in colors with signs of gas production are seen throughout the decomposition process. (Photo taken at University of Tennessee's Anthropologic Research Facility and courtesy of the author.)

carbon monoxide (CO) or cyanide poisoning, while bodies in a very cold temperature can retain a pink color "because the cold inhibits dissociation of oxygen from the hemoglobin" (Clark, Worrell, and Pleuss, 1997).

Several hours after death, changes in color and gas distention may appear in the upper abdomen (Figure 2.9). This is due, in part, to bacteria; the primary source can come from the cecum (a pouch between the small and large intestine). As bacteria begin to reproduce, gas is produced, helping to break down the intestinal walls. The gas will mix with deoxygenated blood (blood without oxygen) and add to the color changes. Colors can result from hydrogen sulfide gas that is produced by microbes in the large intestine. This can combine with iron to produce greens, purples, and then, finally, black. As the intestine fills with gas, it can push the intestine close to the surface of the body, sometimes just under the skin. This can make the skin more translucent showing different colors (Snyder-Sachs, 2001). As time goes on, this coloration can change to a black "smear" that can cover the face, torso, and limbs. The term *marbling* is used when there is a streaking or mixing of colors, often in areas where livor is present (Figure 2.10).

Gas

Already obvious, the product most commonly associated with putrefaction is gas. Actually there are many types of gas that can be produced through bacterial fermentation. A number of different types of bacteria will reproduce and produce hydrogen, carbon dioxide, and methane. Other gases can be produced from the breakdown of different kinds of tissue and/or molecules. The characteristic rotten egg smell of hydrogen sulfide can result following breakdown of proteins that contain sulfur. The role of gases will be discussed later in this chapter and then in Chapter 3.

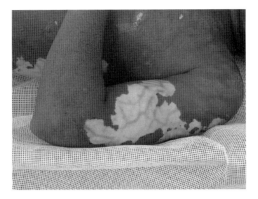

Figure 2.10 Blood caught in blood vessels may become visible through the skin. This is sometimes called "marbling."

Other Products of Putrefaction

Alcohols and acetone also can be produced during the active bacterial growth stage. Because proteins are made up of building blocks called amino acids, when they break down, organic acids are released. The release of these acids can cause a localized increase in pH that can eventually progress to the entire body. As decomposition continues, the acidity plays out and changes to a basic (alkaline) environment. This will be covered more in Chapter 5.

Bloat

A summary by Snyder-Sachs (2001) described the process of bloating. Simply put, because the intestines are no longer moving products along (peristalsis), the bacterial population also stops and becomes more concentrated. Additionally, the microbial populations are no longer controlled by circulating white blood cells, and with no more restraints present, conditions are perfect for a microbial party. The uncontrolled bacteria eat and reproduce, creating gas and other by-products. Filling up the intestine with gas is called active or primary bloat.

As the intestinal cells continue to decompose, they can detach from each other, creating holes in the intestinal wall. This allows bacteria and gas to leak and escape into the chest and abdominal cavities. As bacteria spread and gas production continues, increasing pressure from gas throughout the body can cause the disfigurement that can be seen in some decedents. Gas can fill limbs, genitals, tongue, and lips to many times their normal size.

Gas also can be trapped in the skin and look like localized bubbles or blisters (Figure 2.11). These are called bullae. Like blisters, these bullae can break, dry up, and then lift away from the underlying layers of skin, as can be seen in Figure 2.12. The bullae should not be confused with blisters as bullae occur *after* death. Then, different layers of skin cells can break down

Figure 2.11 Gas trapped in small balloons right under the skin may look like blisters. These are called "bullae."

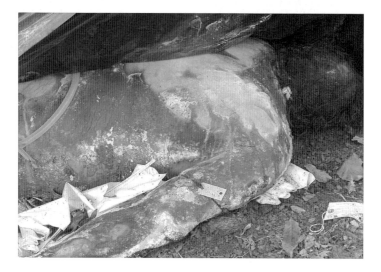

Figure 2.12 As cells break down and pull apart from each other, structures begin to break down as well. One example is skin slippage. In this photo, the skin has stuck to the tarp as it was lifted up. (Photo taken at University of Tennessee's Anthropologic Research Facility and courtesy of the author.)

and detach from each other, resulting in holes in the skin. This can continue to large areas of skin detaching from the body; this is called "skin slippage" (Clark, Worrell, and Pleuss, 1997). A common example of this is when skin of the hand can slide off like a glove.

Stage 3: Post Putrefaction

Liquefaction

Depending on a number of factors, the organs with the greatest blood supply will liquefy first and may begin to seep out of different body orifices. As

Figure 2.13 Frothy foam in the upper part of the photo likely from gas was produced by bacteria in the rectum of a pig.

the organs in the chest and abdomen combine with blood and contents from the abdominal cavity decompose, they develop into a mushy, doughy consistency. This "mush" continues to break down into a dark, smelly liquid called *purge fluid*. As the internal pressure continues to build and the muscle walls that once held it all together break down, purge fluid can be forced out of the nose, mouth, and rectum. Often the abdominal wall will "blow out" causing a flood of liquid from decomposed organs, blood, gas, etc. Purge fluid can mix with gas produced by intestinal microbes, giving it a frothy look (Figure 2.13). The purge fluid can provide a rich food source for bacteria and insects (Figure 2.14).

Liquefaction can occur in days, but, again, the time depends on environmental conditions. Following liquefaction, the body will begin active decay and start to dry out. From this point on, there is less odor produced as the body further breaks down (Rebmann, David, and Sorg, 2000).

Final Breakdown (Destruction and Decay)

One of the last parts of the body to decompose is the musculoskeletal system. Bones are held together by ligaments and cartilage and are connected to muscles with tendons (Figure 2.15). Ligaments are made up of bands or cords of collagen (a protein) that connect two bones together, forming a joint (White and Folkens, 2005). Ligaments, tendons, and cartilage are tough,

Figure 2.14 Large white maggot mass with purge fluid surrounding the decedent. (Photo taken at University of Tennessee's Anthropologic Research Facility and courtesy of the author.)

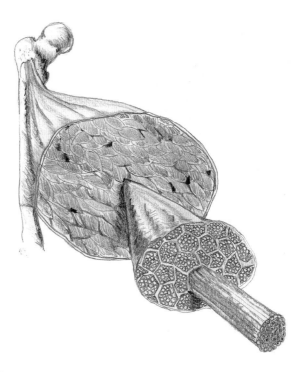

Figure 2.15 Connective tissue holding bones together are some of the last to decompose. (Adapted from Hole, J. W. 1990. *Human anatomy*. Dubuque, IA: Wm. C. Brown Publishing.)

compressed, connective tissue that doesn't have much blood supply. They generally take a long time to break down. Until the ligaments break down, the bones will be more or less held together.

One of the later tissues to decompose is hair. Although there is a difference between animal and human hair, hair generally has a tough outer cuticle (shell) surrounding an inner core (Rowe, 1997). It took six months for human hair buried in "well-watered garden soil" to deteriorate (Rowe, 1997).

In general, live bone contains a dense outer shell surrounding the marrow cavity (Schultz, 1997). As the PMI increases, the marrow (either red with blood or yellow with fat), breaks down and leaks out, giving the bone a greasy look and feel (Lyman and Fox, 1997). With time, the bones can dry out, becoming flaky, brittle, and eventually fall apart (Figure 2.16). The time it takes for bone to reach this stage depends on the age of the decedent and the environment in which the body was placed.

Ross and Cunningham (2011) described the Behrensmeyer's five stages of bone weathering, which are summarized in Table 2.3. The time to progress from stage 0 to 5 "depends heavily on its microenvironment."

Figure 2.16 Skeletonization is the last part of decay. (Photo taken at University of Tennessee's Anthropologic Research Facility and courtesy of the author.)

Table 2.3 Stage of bone decomposition*

Stage	Description
0	Fresh, defleshed bone
1	Cracks in top layer along length of bone
2	Flaking of top layer, soft tissue still present
3	Top layer all gone, deeper layer of compact bone has fibrous quality
4	Overall fibrous appearance and splintering of bone
5	Inner bone exposed, eventually bone falls apart and loses shape

* Adapted from Ross, A. H. and S. L. Cunningham, 2011.

As the body starts to fall apart (but still has bone and connective tissue), it is called *partial skeletonization*. Once the bones are no longer connected, and the skeleton falls apart or disarticulates and is considered to be *fully skeletonized*.

Special States of Decomposition

Different parts of a body will decompose differently, depending on the conditions to which the body is subjected. There are some "special" situations that are a result of certain situations.

Mummification

Mummification is pretty straightforward and is simply when tissues slowly dry out through evaporation of fluid. There can be partial or complete mummification, and it can happen in extremely hot or cold situations (Figure 2.17).

True or complete mummification generally occurs only in very hot or very cold environments. In ancient times, Egyptian mummies were treated with spices and herbs, then wrapped and put in a sarcophagus. This resulted in complete mummification. However, extreme cold also can cause preservation of bodies. There are many examples of mummified bodies found in snow or ice packs. In either situation, the body virtually dries out.

Partial mummification is more common than most people think. When a body is exposed to wind or hot, dry conditions, parts of the body can slowly

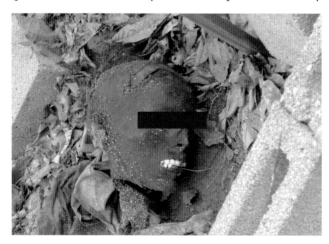

Figure 2.17 Extreme dryness (either in hot or cold conditions) can cause total or partial mummification. (Photo taken at University of Tennessee's Anthropologic Research Facility and courtesy of the author.)

dry out causing the skin to look like leather stretched over bone. This can be seen in parts of the body that are not covered or protected by clothing.

Adipocere: Grave Wax or Death Soap

Adipocere is what can happen when body fat (adipose) breaks down under special conditions, usually when the body is in a wet area where there is a lack of oxygen (O_2) available (Figure 2.18).

There is an example that may help explain how adipocere can form. When people began to settle this country, they didn't have the luxury of store-bought soap and were forced to make their own. To do this, they used either tallow (fat from cattle, deer, or sheep) or lard (fat from pigs). Usually they made soap after butchering a pig. The process of making soap is called saponification and is what happens when a fatty acid meets an alkali (base) in a watery environment (Anon., n.d.). In those times, the alkali would come from wood ash lye (potassium). Lye would be mixed with cleaned fat in water that was then heated to boiling, usually in a big kettle outside over a fire. After about six to eight hours of boiling, the fat and lye would form together to make soap.

The process of making soap is similar to what happens when adipocere is formed. It takes fat, then bacteria in an environment with lots of moisture and a warm temperature. There is one specific type called *Clostridium perfringens* that is commonly found both in the intestine and soil (Forbes, 2008). This anaerobic bacteria (along with others) prefer to live in areas where there are low levels of O_2. *C. perfringens* bacteria produce enzymes that breaks down protein

Figure 2.18 Adipocere occurs from breakdown of fat in low oxygen and moist conditions. (Photo taken at University of Tennessee's Anthropologic Research Facility and courtesy of the author.)

Figure 2.19 Textures of adipocere vary depending on environmental condition, much like textures of various cheeses. (Photo by J. K. Rosenbaum.)

and fats, supplying nutrients. This is the process of fermentation and, as part of the fermentation process, glycerol (an alcohol) is produced (Forbes, 2008). The presence of water and the acidic environment causes the glycerol to further break down into free fatty acids (FFA). If conditions continue to be right (low O_2, high water), the bacteria breaks down body fat until it is a "mass of fatty acids" (Forbes, 2008). When in a neutral or slightly alkaline area, the FFA can combine with sodium and/or potassium minerals from the body. This creates a whitish, smelly, crumbly substance, resembling feta cheese.

When the body is buried, the sodium and/or potassium may be removed and replaced with calcium or magnesium found in the soil or water. If this happens, adipocere turns into something like a hard, shredded cheese (Figure 2.19). An interesting point: The regular soap made by settlers was a soft soap made from lye and lard. To make the soap hard, they added salt to the last part of the boiling process so the adipocere's potassium would be replaced with sodium from the salt. As adipocere ages, it can dry out and become hard and brittle. Adipocere has been known to encase an entire body, delaying or stopping further decomposition (Judah, 2008).

Some other interesting facts about adipocere:

- If there are warm, moist conditions, adipocere can begin to form within a few days, but usually takes up to several weeks.
- Adipocere may be more evident in females and/or infants due to their higher fat content. It can be found in parts of the body with high fat content.
- The high water content in fat cells (adipocytes) may offset the lack of moisture in a dry environment, or an obese person may form adipocere even if they are not near water.
- Adipocere can last longer in submerged bodies.

- Adipocere and mummification can occur in the same decedent; areas covered with clothing enhance formation of adipocere (Clark, Worrell, and Pleuss, 1997).

Summary

The following may provide an easier "Rule of Thumb for Time of Death" (Snyder-Sachs, 2001).

> If warm and not stiff: Dead for a few hours
> If warm and stiff: Dead for less than 12 hours
> If cold and stiff: Dead for 12–48 hours
> If cold and not stiff: Dead for more than 48 hours

Obviously, decomposition is completely dependent on the environmental conditions that the body is in at the time of death. Stay tuned for more information in Chapter 3.

References

Anon. n.d. Lard for making soap: Both lard and castile soap are old fashioned. Online at: www.natural-soap-making.com/lard-for-making-soap.html (accessed July 6, 2011).

Carter, D. O., and M. Tibbett, eds. 2008. Cadaver decomposition and soil: processes. In *Soil analyses in forensic taphonomy: Chemical and biological effects of buried human remains.* Boca Raton, FL: CRC Press, Chap. 2.

Clark, M. S., M. B. Worrell, and J. E. Pleuss. 1997. Postmortem changes in soft tissue. In *Forensic taphonomy: The postmortem fate of human remains,* ed. W. D. Haglund and M. H. Sorg. Boca Raton, FL: CRC Press, Chap. 9.

DeGreeff, L. E., and K. G. Furton. 2011. Collection and identification of human remains volatiles by non-contact, dynamic airflow sampling, and SPME-GC/MS using various sorbent materials. *Analytical and Bioanalytical Chemistry* 401(4):1295–1307.

Forbes, S. L. 2008. Decomposition chemistry in a burial environment. In *Soil analysis in forensic taphonomy: Chemistry and biological effects of buried human remains,* ed. M. Tibbett and D. O. Carter. Boca Raton, FL: CRC Press, Chap. 8.

Haglund, W. D., and M. H. Sorg, eds. 1997. Method and theory of forensic taphonomic research. In *Forensic taphonomy: The postmortem fate of human remains.* Boca Raton, FL: CRC Press.

Hole, J. W. 1990. *Human anatomy and physiology.* Dubuque, IA: Wm. C. Brown Publishing.

Judah, J. C. 2008. *Buzzards and butterflies: Human remains detection dogs.* Brunswick County, NC: Coastal Books Publishing.

Lyman, R. L., and G. L. Fox. 1997. A critical evaluation of bone weathering as an indication of bone assemblage formation. In *Forensic taphonomy: The postmortem fate of human remains,* ed. W. D. Haglund and M. H. Sorg. Boca Raton, FL: CRC Press, Chap. 15.

Merriam-Webster Dictionary. Online at: www.merriam-webster.com (accessed August 23, 2011).

Rebmann, A., E. David, and M. H. Sorg. 2000. *Cadaver dog handbook: Forensic training and tactics for the recovery of human remains.* Boca Raton, FL: CRC Press.

Ross, A. H., and S. L. Cunningham. 2011. Time since death and bone weathering in a tropical environment. *Forensic Science International* 204: 126–133.

Rowe, W. F. 1997. Biodegradation of hairs and fibers. In *Forensic taphonomy: The postmortem fate of human remains,* ed. W. D. Haglund and M. H. Sorg. Boca Raton, FL: CRC Press, Chap 21.

Schultz, M. 1997. Microscopic structure of bone. In *forensic taphonomy: The postmortem fate of human remains,* ed. W. D. Haglund and M. H. Sorg. Boca Raton, FL: CRC Press, Chap. 13.

Snyder-Sachs, J. 2001. *Corpse: Nature, forensics and the struggle to pinpoint time of death.* Cambridge, MA: Persuas Publishing.

White, T. D., and P. A. Folkens. 2005. *The human bone manual.* New York: Elsevier Academic Press.

Bugs, Bodies, and Beyond

3

Introduction

This chapter covers how the environment affects decomposition of the body and is not all inconclusive, but hits the high points of primary and secondary variables that affect decomposition. As there are many things that affect what taphonomic changes may be seen, this is a rather extensive chapter. The effects of environmental variables will be a recurring theme throughout the rest of the book. In addition to temperature, there are brief summaries of the effects of soil types (particularly in reference to buried decedents) and water (in reference to decedents on land and in water). Chapter 4 will cover the effects of the environment on odor dispersion.

Primary Environmental Factors

As discussed in Chapter 2, primary factors affecting decomposition are temperature, water, and energy. After death, the energy in the body dissipates. The effect of water, or lack thereof, has already been addressed with adipocere formation or mummification. Water as a secondary factor (drowning victims) will be covered later. So, to simplify things, it comes down to temperature.

Temperature

High temperature relates to high soil temperature, high levels of soil bacteria, lots of insects, and warm water. Conversely, low temperatures result in lower soil temperature, less bacteria, less insects, and less water. Remember, high temperature also causes accelerated chemical reactions and decomposition, while cold results in slower decomposition.

Therefore, temperature can refer to ambient (air) temperature, soil and/or water temperature, body temperature, and cellular temperature. Temperature controls the rate of the chemical reactions in the cells, both in the live and dead body. Temperature controls the amount and type of bugs that might be present. Bugs include all types and sizes, e.g., flies, beetles, spiders, protozoa, bacteria, etc.

To help understand the differences in the type and rate of decomposition, it helps to "standardize" time based on Accumulated Degree Days or ADD. The definition of ADD varies depending on what field a researcher is in (Hadley, n.d.). In agriculture, degree days may mean the temperature required for a certain plant to germinate after being planted; in entomology, degree days may refer to temperature that is required for a fly egg to hatch into a maggot and grow. The ADD may vary with each type of plant or type of insect (Catts and Haskell, 1990). The word *accumulated* means that the average daily temperature is added together from one day to another. Michaud and Moreau (2011) studied the use of a degree-day approach to study decomposition-related processes and found that ADD can be used as a scientific decompositional baseline to estimate the postmortem interval (PMI).

Burial accumulated degree days (BADD) is the average temperature in a grave taken each day and then added together over time.

We will next take a closer look at the reasons why there are differences in decomposition, both in rate and how it takes place. Along with temperature, one of the biggest contributors to decomposition is bugs. This is where we look a little closer at the role that bugs play.

Microbiology: The Little Bugs

The intent of this section is not to dive deeply into microbiology, but instead to provide enough background to understand the role of "little bugs" on decomposition. One of the major sources of microbes is the body itself, described as having a "heavy microbial inoculum in the form of enteric and dermal microbial communities" (Carter, Yellowlees, and Tibbett, 2008), meaning the bugs are found mainly in the intestinal tract and on the skin. Depending on the ambient (air) and soil temperature, soil microbes also can help with decomposition if the body is in contact with the soil surface or in a shallow grave. Different types of soil and outside areas will have different types and number of microbes (Figure 3.1). This can cause obvious differences in decomposition (Metting, 1993).

Some basics about little bugs include (Gill-King, 1997):

- As the body's cells die, O_2 (oxygen gas) becomes depleted.
- Decreasing O_2 levels support rapid multiplication of anaerobic intestinal (especially large intestine) bacteria as well as soil microbes.
- That bacteria start to "operate quickly on host cells in their immediate environment" breaking down carbohydrates, proteins, and fats that cause the production and build up of gases.

More of this will be covered in Chapter 4.

Figure 3.1 Soil type can greatly affect the decomposition process.

Entomology: The Big Bugs

Medicocriminal entomology is a branch of science whose role "is to examine and identify arthropods on or near a corpse" and then use that information to draw conclusions to determine accurate estimates relative to the interval of time that a body has been exposed to arthropod activity (Catts and Haskell, 1990; Haskell et al., 1997). Much of the information in this section comes from those two references. An important point is that the focus of forensic entomology is *not* to determine time of death, but instead to determine how long a body had been exposed to bugs (Figure 3.2).

Entomology or the study of "big bugs" (insects) starts with a brief overview of arthropods or critters with jointed legs (Catts and Haskell, 1990). The type of bugs that are pertinent to decomposition can be generally broken down to insects (six legs, including flies), arachnids (eight legs, including spiders and mites), and multiple-legged creatures (including centipedes and millipedes). It is important to point out that arthropods play a huge role in decomposition *and* are extremely useful in helping to determine how long a decedent has been in that environment with bugs (Figure 3.3). The following is a 10,000-foot overview of some of the major types of arthropods associated with decomposition.

Flies

The insect most commonly associated with decomposition is the fly. The basic life cycle of the fly is:

Figure 3.2 "Big bugs" or arthropods also affect how a body breaks down.

- Egg
- Larvae ("instar" stages, the last commonly called maggot)
- Pupae
- Adult (Figure 3.3).

According to Catts and Haskell (1990), "... environmental factors such as temperature, day length, available food supply, and humidity may serve to limit the times of the year during which insects can develop. The species and forms of insects present on or under the corpse may vary with the geographic location and the time of the year."

Under the "right" conditions, flies are normally the first insects to come to a body, reportedly as early as minutes after death (Catts and Haskell, 1990). Flies will lay eggs in cavernous-like openings in the body that provide a dark, moist place where the eggs can hatch into larvae or maggots and grow (Figure 3.4). The maggots feed on decomposing remains and are "intimately associated with decomposition." It is well recognized that insects can break down a body to bones within a few weeks if the conditions are right. With "ideal" conditions, fly eggs hatch within a few hours (6–40 hours) go through molts, and then crawl off to pupate for 6–18 days (Pope, 2010). Flies play a major role in determining the length of time a body has been in a specific environment.

There are many types of flies associated with remains; each type arrives at different times, depending on weather and temperature conditions. The types of flies will vary depending on geographic location, therefore, the

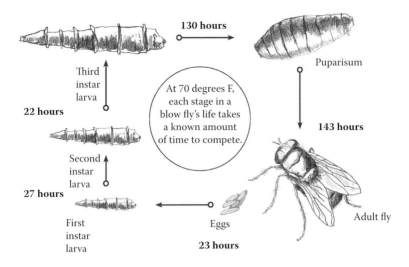

Figure 3.3 The basic fly life cycle from egg to larvae or maggot (usually three stages) to adult. The time will vary with the type of fly. (From Google image. Online at www.nlm.nkh.gov/visibleproofs/media/detailed/ii_a_216n.jpg)

Figure 3.4 Flies like to lay eggs in dark, moist areas out of direct sunlight, especially nostrils.

scientific names (genus and species) will not be used here. There are several references that can be consulted if more information is needed. For this chapter, however, the flies will be unceremoniously grouped into blowflies, flesh flies, and house flies.

The number of stages and the time it takes to move through each of them depends on the species of flies and environmental factors, especially temperature. In cool conditions, **blow flies** arrive early and lay eggs around any natural body opening accessible to them. This is generally around or in the eyes, ears, mouth, anus, and any open wound that may be present. If you think about it, eggs and larvae around the orifices makes sense (Figure 3.5). The cave-like openings provide a nice, dark, and usually moist place to lay eggs to incubate and then adequate fluids in which to grow. As the tissues continue to break down, the fly larvae go through several life stages, feeding and breaking down more tissue, developing into the maggots that we all recognize. The final stage maggots can form what is called a "maggot mass," a writhing, flesh-eating, heat-producing, massive mob of destruction that, in the right conditions, can literally decimate a body in a matter of days (Figure 3.6). **Flesh flies** also arrive early, again with eggs laid and larvae being commonly present around openings. Later the **house flies** arrive, all adding

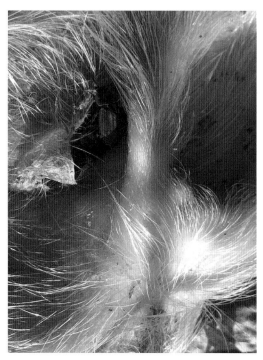

Figure 3.5 Natural orifices in the body are a desirable place for flies to lay eggs and for beetles to crawl into.

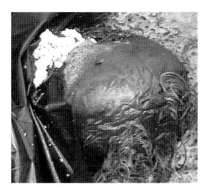

Figure 3.6 When the eggs hatch, they can form maggot masses (a large white mass). These masses can give off a great deal of heat and can decimate a body in a matter of days. (Photo taken at University of Tennessee's Anthropologic Research Facility and courtesy of the author.)

to the reproducing and feasting maggots. In warmer conditions, flesh flies usually are among the early guests, but blow flies and house flies will be present as well. In the early stages, the adult flies feed on decomposition fluids.

Flies will continue to lay eggs that hatch and complete the circle of life in and around a decomposing body. As the maggots eat and grow, they eventually will crawl away from the body, find a nice dark place (usually under the body or in the soil), and then form into a pupa. The pupal casing of the soon-to-be adult fly is encased in a brown-colored cocoon that darkens over time. Pupal casings can often be found several feet away from the body, often in the soil. Remember that these may be found *after* the body has been removed from the area.

Other Bugs

Many species of beetles also are often associated with decomposing remains. Some of those beetles come in within a few hours and will remain through the later stages of decay (Figure 3.7). Others come in later because they are predators, and feast on the larval and (sometimes) adult flies. Some of the common beetles include:

- *Dermestid (skin) beetles*: Adults and larvae feed on dried skin and tissue. They serve as a source of food for other beetles.
- *Carrion or burying beetles*: Associated with early decay and remain until the dry stage. They feed on fly maggots and adults.
- *Hister beetles*: Feed on maggots.
- *Scarab beetles*: Usually associated with dung and plant decay; they are scavengers and predators of larvae and adults.

Figure 3.7 Depending on conditions, beetles of many types will assist during the decay process.

Cockroaches are not commonly associated with fresh decomposition, but being omnivorous, they will feed on decomposition fluid and tissues during the later stages. Earwigs also may be found in high numbers in and around decomposing remains.

Then more bugs arrive, including:

- *Ants*: Are recognized predators that eat fly maggots. Ants can actually cause a significant decrease in the number of maggots and slow decomposition.
- *Pillbugs*: Eat rotting vegetation and carrion and are frequently seen throughout decomposition.
- *Spiders*: Do not feed on carrion, but are predators of other arthropods.
- *Mites*: Can come in attached to flies (transport service?) or from the soil under the body. Mites feed on eggs and small larvae and, later in decay, may feed on fungus that grows under the remains.
- *Centipedes/millipedes*: Are frequently found around remains and will feed on both remains and plants in the area. Millipedes are generally found in moist, protected areas.
- *Bees and wasps*: May feed on remains; some lay eggs near remains.

Other wingless bugs (including springtails and silverfish) that like damp and dark areas may eat remains as well. Even butterflies (and moths) are attracted to decomposing tissue and fluids. According to Judah (2008), the "Lesser Purple Emperor [butterfly] can smell a body from hundreds of meters away." Judah describes how butterflies that normally group in wet areas will instead congregate and feed at a carcass or nearby soil wet with purge fluid.

To review, in "ideal" conditions, flies arrive and lay eggs, and maggot masses grow and feed as the body progresses to the bloated stage. Time marches on and gases are released, the abdomen decompresses, and fluid seeps out. At this point, flies continue to reproduce and feed, then carrion and clown beetles come in and help out. Mold may form and attract other types of bugs. During the dry stage, beetles help with removal of remaining tissue, along with other scavengers.

Now that temperature and bugs have been addressed, it is time to examine the actual remains . . . of pigs.

Meet the Pigs

This part includes a series of photographs that were taken over time of two pigs that were placed in the same general area in the central Midwestern United States. One pig died of natural causes and the other was euthanized after prolapsing his rectum. Both were donated for teaching purposes. Each carcass was placed in a wire dog crate in direct contact with the ground. The crate was placed on the edge of a forest bordering a cultivated farm field and was used to protect the carcass from an established coyote population. There was no evidence of predation during the times either pig carcass was out in the open. The intention of these photographs is to demonstrate the difference that temperature can make in the rate and course of decomposition on bodies deposited on the surface of the ground.

The first pig ("Hank, the Winter Hog") was placed out in mid-February 2010; the second ("Horace, the Summer Hog") was put out in mid-June. The photographs are organized based on temperature. For the pig part, ADD will be described as the sum of all daily mean temperatures above 0°C that "comprise the postmortem interval time span" (Megyesi et al., 2005). Often 5–10°C is subtracted from the daily temperature to account for the minimum development threshold for insects. For this exercise, the average temperature from the closest "official" weather station was collected, converted to Celsius scale, and any temperatures over 0°C were added from the time the pigs were placed outside to the point the photographs were taken. Table 3.1 shows it all (Figure 3.8–Figure 3.67).

Table 3.1   The pigs

Figure No.	Date	Postmortem Interval	Accumulated Degree Days
Hank, the Winter Hog			
3.8	2/17	D0	0
3.9	2/17		
3.10	02/27	D10	128.5°F/53.6°C
3.11	03/07	D18	206°F/97°C
3.12	03/07		
3.13	03/07		
3.14	03/18	D29	606°F/319°C
3.15	03/27	D38	852°F/456°C
3.16	04/03	D45	1040°F/560°C
3.17	04/09	D51	1305°F/707°C
3.18	04/09		
3.19	04/09		
3.20	04/21	D63	1882°F/1028°C
3.21	04/21		
3.22	04/21		
3.23	5/1	D73	2390°F/1310°C
3.24	5/1		
3.25	5/16	D88	3268°F/1798°C
3.26	5/16		
3.27	5/16		
3.28	5/16		
3.29	5/16		
3.30	5/16		
3.31	5/31	D103	4176°F/2302°C
3.32	5/31		
3.33	5/31		
3.34	5/31		
3.35	5/31		
3.36	10/1	~7.5 months	
3.37	10/1		
3.38	10/1		
Horace, the Summer Hog			
3.39	6/9	D0	0
3.40	6/9		
3.41	6/11	D2	134°F/57°C
3.42	6/11		
3.43	6/11		
3.44	6/11		
3.45	6/11		

(continued)

Table 3.1   The pigs (continued)

Figure No.	Date	Postmortem Interval	Accumulated Degree Days
3.46	6/12	D3	
3.47	6/12		
3.48	6/13	D4	257°F/130°C
3.49	6/13		
3.50	6/13		
3.51	6/13		
3.52	6/13		
3.53	6/13		
3.54	6/16	D7	448°F/231°C
3.55	6/19	D9	659°F/348°C
3.56	6/19		
3.57	6/19		
3.58	6/19		
3.59	6/19		
3.60	6/19		
3.61	6/19		
3.62	6/22	D13	953°F/512°C
3.63	6/22		
3.64	6/22		
3.65	6/22		
3.66	10/1	~4 months	
3.67	10/1		

Figure 3.8 Hank the hog delivered to site, mid-February.

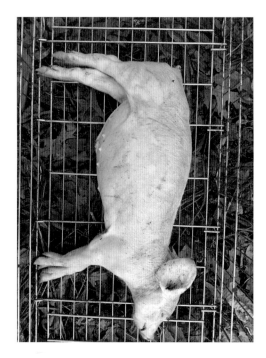

Figure 3.9 Hank in crate, dead less than 12 hours. Post Mortem Interval (PMI) Day 0; 0°F.

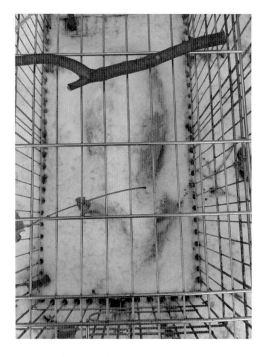

Figure 3.10 Snow-covered Hank, freeze and thaw. PMID 0; 129°F.

Figure 3.11 Hank with some bloating and eyes retracted. PMID 18; 206°F.

Figure 3.12 Hank's nose drying out and eyeballs receded. PMID 18; 206°F.

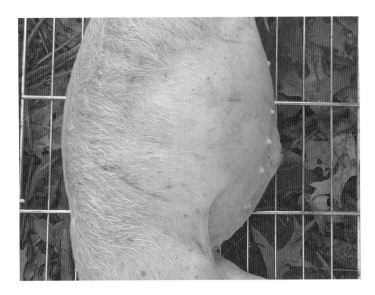

Figure 3.13 Abdominal bloat and thinning skin. PMID 18; 206°F.

Figure 3.14 Minimal evidence of fluid leakage. PMID 29; 606°F.

Figure 3.15 Drying of skin (placed directly on soil). Slight increase in abdominal bloat. PMID 38; 852°F.

Figure 3.16 Beetles begin to appear. PMID 45; 1040°F.

Figure 3.17 Drying continues, beetles attack eyes and ears. PMID 51; 1305°F.

Figure 3.18 Hank from the side, lots of beetles. PMID 51; 1305°F.

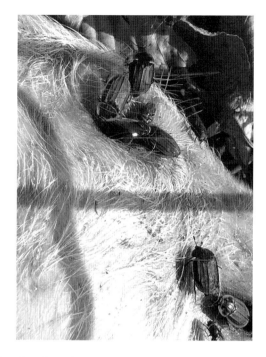

Figure 3.19 More beetle infestation. PMID 51; 1305°F.

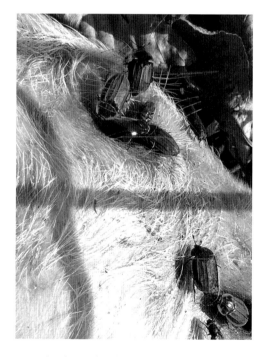

Figure 3.20 Seven weeks from death—beetles continue. 1305°F.

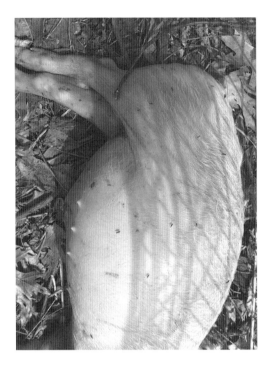

Figure 3.21 Hank's side, bloating continues. PMID 63; 1882°F.

Figure 3.22 The beetles continue. PMID 63; 1882°F.

Figure 3.23 Notable skin breakdown. PMID 73; 2390°F.

Figure 3.24 Hank—intestinal microbial "parties" creating gas. PMID 73; 2390°F.

Figure 3.25 Hank—gas leaves and the abdomen collapses. PMID 88; 3268°F.

Figure 3.26 Hank—front legs down to bone. Beetles! PMID 88; 3268°F.

Figure 3.27 The beetles continue on the face. PMID 88; 3268°F.

Figure 3.28 Closer view of collapsed abdomen. Skin appears thickened and beetles at work. PMID 88; 3268°F.

Figure 3.29 The beetles at work on rear legs. PMID 88; 3268°F.

Figure 3.30 A few flies join in. PMID 88; 3268°F.

Figure 3.31 Two weeks later—some skin, lots of bones, and lots of beetles. PMID 103; 4176°F.

Figure 3.32 The whole Hank (head at top). PMID 103; 4176°F.

Figure 3.33 Hank—head. PMID 103; 4176°F.

Figure 3.34 Hank's ribs coverered with hair; purge fluid with beetles on center. PMID 103; 4176°F.

Figure 3.35 Hank's rear legs (and beetles). PMID 103; 4176°F.

Figure 3.36 7½ months later

Figure 3.37 The cleaning job—done by the beetles (7½ months later).

Figure 3.38 The big bug break!

Figure 3.39 Horace the summer hog in mid-June.

Figure 3.40 Horace on the ground. PMID 0.

Figure 3.41 Sawdust-like fly eggs in ears, nose, and the corner of the mouth. PMID 2; 134°F.

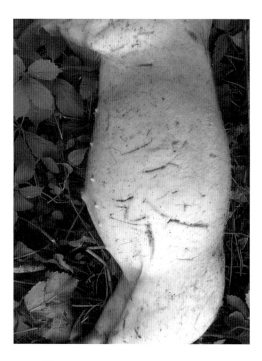

Figure 3.42 PMID 2—bloat. 134°F

Figure 3.43 Prolapsed rectum attracts flies. PMID 2; 134°F.

Figure 3.44 Fly eggs and stuff starting around the mouth. PMID 2; 134°F.

Figure 3.45 Clusters of white fly eggs on the head. PMID 2; 134°F.

Figure 3.46 Day 3—bloat.

Figure 3.47 PMID 3—the attack of the flies.

Figure 3.48 Day 4—face covered with fly larvae (baby maggots) and more eggs. PMID 4; 257°F.

Figure 3.49 Close-up of face and maggots. PMID 4; 257°F.

Figure 3.50 Ears and maggots. PMID 4; 257°F.

Figure 3.51 Whole pig bloat. PMID 4; 257°F.

Figure 3.52 Prolapsed rectum—a popular area for flies (maggots). PMID 4; 257°F.

Figure 3.53 First showing of foam—probably gas produced by rectal bacteria. PMID 4; 257°F.

Figure 3.54 PMID 7—major foam on rear quarters, plus maggots, etc. 448°F.

Figure 3.55 Day 9—whole pig turning into skin with bones. 659°F.

Figure 3.56 Bony face with skin. PMID 9; 659°F.

Figure 3.57 D9—rear legs with LOTS of maggots.

Figure 3.58 Fly maggots—growing up! On skin and soil. PMID 9; 659°F.

Figure 3.59 Rear feet with maggots under skin. PMID 9; 659°F.

Figure 3.60 Feet, with maggots traveling from body. PMID 9; 659°F.

Figure 3.61 Dried skin over skull. Flies working "undercover" destroyed tissue under skin. PMID 9; 659°F.

Figure 3.62 Day 13—the whole hog. 953°F.

Figure 3.63 Dry leathery skin covering the face. PMID 13.

Figure 3.64 D13—leathery dry skin covering the front legs. 953°F.

Figure 3.65 Rear legs on PMID 13. Skin covering the bone. 953°F

Figure 3.66 Four months later—dry skin and bones.

Figure 3.67 Four months later—Horace is skin and bones, not white, not clean (few to no beetles?).

Observations

So what does this all mean? Table 3.2 summarizes the differences between the environmental factors and the effects on decomposition of Hank and Horace. To put this into perspective, Vass et al. (1992) reported that a 68 kg human progressed to advanced decay at an ADD between 400 and 1285°C. In a review and study conducted by Pope (2010), Table 3.3 summarizes the stage of decomposition with postmortem interval (PMI) days and ADD (°C) of bodies found outside on the soil surface.

Megyesi (2005) studied the predictability of PMI and ADD and found that ADD accounted for about 80% of the variation in decomposition.

So, what can be determined by these studies? Hank's ADD for skeletonization fell within the Pope range, but Horace's did not. Although ADD is a way to track the decomposition process, in temperate climates, decomposition is *very* dependent on the contributions of bugs. The end result of skeletonization is ultimately the same, but there are profoundly different ways to get there, depending on temperature, arthropods, and microbes.

Table 3.2   Summary of differences between Hank and Horace

Observations	Hank	Horace
Total ADD	4176°F/2302°C	953°F/512°C
Date out	Feb 12	June 9
Date to bone	May 31	June 22
Postmortem interval (to bone)	103 days	13 days
Physical differences	• Some bloat early due to exposure to ~80°F temperature for 12 hours after death, then moved outside • Frozen • Thawed • Once warmed, some minor bloating, few flies, and many beetles • Dried out evenly • Bones with some hair • Decomposition from outside in; probably beetles and soil bugs primary influence	• Euthanized due to prolapsed rectum • Out in warm weather • Many flies and eggs by Day 2 • Pronounced bloat and fly eggs and maggots by Day 4, especially on anus and face • Profound bloat and maggots followed by liquefaction and more flies and maggots • Skin left somewhat intact covering bones • Decomposition from the inside out; big and little bugs working in concert with each other.

Table 3.3   Stages of decomposition, postmortem intervals, and accumulated degree days[*]

Stage	PMI Days	ADD (°C)
Fresh	1–59	0–818
Bloated	1–102	0–342
Advanced	2–76	9–1144
Decay	135–4026	1800–2168

[*] Adapted from Pope, 2010.

An Interesting Pig Story

One researcher described a study done in a suburban area in Canada in which three pigs were placed outside and three others placed inside in separate rooms of a house (Anderson, 2011). The outside ambient temperature ranged from 39.2–104.4°F (4 to 43°C) over the 42-day period, with an average daily temperature of 61.5°F (16.4°C). Inside temperatures ranged from 53–104°F (11.5 to 40°C), with an average daily temperature of 64°F (17.8°C). As one would expect, there was a great deal of insect colonization (eggs, maggots, flies, etc.) on the outside pigs, causing a much faster rate of decomposition and time to skeletonize. There was a delay of five days for colonization for the inside pigs. This was not a big surprise. This chapter provides a very detailed description of the pigs' taphonomic changes along with the entomological timelines.

Secondary Environmental Variables

After the graphic and explanatory pig pictures, it's now time to consider other things that can affect decomposition. According to Casper's law, the speed of decomposition to mummification or skeletonization depends on the body having free access to air. If exposed to free air, the body decomposes twice as fast as a body submerged in water and eight times as fast as a body buried in soil (Judah, 2008). The location of the body can obviously affect the rate and stages of decomposition.

Soil

Location, location, location—it's everything to realtors *and* taphonomists. A big part of that location is soil. Without becoming a pedologist (a soil scientist), there are some basics that may be helpful in understanding how soil can affect decomposition of a body, located either on the surface or buried underneath. Much of the information in this section is from *Soil Analysis in Forensic Taphonomy* (2008) edited by M. Tibbett and D. O. Carter. Some of the information is also from W. C. Rodriguez (1997).

So, here's the dirt on dirt. Dirt is not the same as soil, but soil can be dirt. Whatever word is used to describe it, different professions look at it differently. To the pedologist, soil is made up of different sized mineral particles and organic materials and has complex "biological, chemical, physical, and mineralogical properties" that change over time (Fitzpatrick, 2008). To farmers, soil provides a physical and chemical setting for plants that require exchange of gases, nutrients, and water to grow. To the taphonomist, soil can greatly influence the state of decomposition of a decedent and as an HRD dog handler or investigator, it can influence the ability to locate the decedent.

One of the most important points to understand is that not all soils are alike. This seems pretty simple, but there are many different types of soil that have different characteristics. It's important to think of soil on a spatiotemporal basis, that is, it varies by location, climate, parent materials, topography, and soil organisms. Also, the state of soil does not stay the same; it changes over time either from natural reasons or from "anthropogenic" or manmade influences like plowing or digging. Soil differs in layers, type, texture and porosity, consistency, and microbial concentration.

Figure 3.68 Soil differs in many ways. (Photo courtesy Andrew Rosenbaum.)

Layers

Topsoil and subsoil are the ways to describe soil depth (Figure 3.68). Topsoil is usually the top 12–18 inches and is the layer that can be changed by planting, plowing, application of chemicals (fertilizers, bodies, etc.), and traffic. Subsoil is anything below the topsoil (>18 inches deep) that is not modified or changed except by deep excavation or the movement or drainage of elements from the topsoil into the subsoil.

Types

There are many kinds of soil differentiated by the size or diameter of the mineral particles in the soil. The particle size also will affect the porosity of the soil—the larger the particles, the bigger the space between them. It affects the amount of water it holds and how it travels through the soil layers or up to the soil surface. Water moves through soil by a number of different mechanisms including surface runoff, infiltration, storage, and deep drainage. Soil can act as a sink and filter for water. These concepts will become more important in Chapter 5.

Texture and Porosity

Sand is larger than silt, which is larger than clay. The actual particle diameters ranges are: sand, 2–0.02 millimeters (mm); silt, 0.02–0.0002 mm; and clay, < 0.0002 mm (Fitzpatrick, 2008). Table 3.4 contrasts some of the basic differences between sand and clay.

In soil maps (covered in Chapter 6), soil can be described as sand, sandy loam, loam, clay loam, light clay, medium-heavy clay, and heavy clay. Sand also can be described as sand, loamy sand, or clayey sand.

Table 3.4   Characteristics of soil*

Characteristics	Sand	Clay
Porosity	Lots of space between large particles	Compacted, not much space between small particles.
Water and nutrient retention	Not much retained; leaches down or moves up	High retention
Particle diameter (mm)[a]	0.25–2.00 (medium–course)	<0.002
Number of particles/ gram soil[a]	90–5700	90,300,000,000

* Adapted from Alexander, M. 1997. *Introduction to soil microbiology*, 2nd ed. New York: John Wiley & Sons.

Consistency

Soil can be described by its consistency or how well the soil holds together. If you squeeze a handful, will it hold together or fall apart? It can be loose, soft, firm, or rigid. This, obviously, will depend on the type of soil and the water content.

Microbial Concentration

As already mentioned, soil also differs by the microbial population it may contain (Metting, 1993). This population can change as the soil is disturbed by digging or excavation. Soil also contains free enzymes that can affect chemical reactions. These enzymes are often released by plants or microbes and likely have a very localized effect.

Burial

When a body is buried, decomposition is slowed. The degree of delay will depend upon a number of factors, most likely soil temperature, soil density (type), and depth.

Within the first several inches (<12) of the topsoil layer, the temperature is usually pretty close to the ambient temperature. At these depths, the odors from the body may reach the soil surface and attract arthropods (Pope, 2010). As the odor reaches the surface, blowflies and flesh flies are attracted and lay eggs that hatch on the soil. If the soil is not too compacted, the maggots may be able to work their way down to the body to help with the decomposition process.

At a mid-level depth (about 2 feet), the soil temperature is cooler and body microbes may be less active. Sometimes, decomposition is aided by the action of soil microbes and nematodes (aka worms). There has been evidence of flies traveling down to depths of 2 feet to reach buried remains (E. Pastula, personal communication, 2011).

At depths of around 4 feet, the soil temperature eventually stabilizes and usually does not freeze. In northern regions, water pipes are usually buried at about 4 feet to protect from freezing. If buried deeply (generally at temperatures < 4°C or 39.2°F), there is a lack of putrefaction probably due to a decrease in microbial participation from the body and soil (Pope, 2010). This can actually result in desiccation or drying of the body. Deeper graves may be associated with the formation of adipocere due to lower oxygen conditions and higher moisture.

According to Brookes (2008), the soil microbial mass (basically the amount and concentration of microbes in soil) can make up 1–3% of the total soil organic matter. The more available soil microbes there are, the faster the body may decompose. Decomposition can cause the soil to become acidic around the body, which can act as a fertilizer and help plants grow. There may be disruption to the plants at the burial site, but, if in the summer, the plants may have an opportunity to regrow in the area (Pope, 2010). Plant roots will grow toward the body to pull nutrients out of the soil and into the plant. In long-term conditions, the plant roots can actually wear grooves down into aged bones as they extract minerals for growth.

According to one study, moisture was a dominant parameter affecting decomposition of bodies buried in soil (Carter, Yellowlees, and Tibbett, 2010). Decomposition was faster in loamy sand and sandy soil than in clays. Greater decomposition occurs in wet sandy soil than in wet "fine textured" soil, probably because the fine soil has less microbes and less enzyme activity.

A study was done to evaluate the differences in decomposition of pigs buried in three different field sites in the United Kingdom (Wilson et al., 2007). It included burials in pasture, moorland (field), and deciduous woods. The bodies were checked at 6, 12, and 24 months after burial and samples were collected and analyzed. Results showed that different soils had a marked effect on the condition of the buried bodies, even when they were in the same basic vicinity. This reflects how much variation there is in soil. The authors pointed out that decomposition caused remarkable changes in the local soil as purge fluid was released, changing the soil pH, number of microbes, moisture, and changes in the oxygen levels. The study highlights the importance "of site-specific environmental information" in understanding the taphonomic changes that may be seen.

Decedent: Size and Age

The size and age of the decedent plays a role in decomposition. The bones of juveniles or infants may only be partially calcified and can break down faster than adult bones (White and Folkens, 2005). Smaller bodies often have both a higher fat and moisture content, which may be involved in the formation of adipocere. Smaller bodies are also easier for scavengers and predators to attack and scatter (Morton and Lord, 2006). Bodies of large mass are not as likely to be scavenged, but body size was found to be a significant factor when insects had access (Simmons, Adlam, and Moffatt, 2010).

Decedent: Clothing

One study looked at the effects of clothing and/or wrapping a body on decomposition and arthropod infestation (Kelly, VanderLinde, and Anderson, 2009). In this experiment, eight pigs were killed and immediately placed outside, some in the summer and some in the autumn. Some of the pigs were dressed in tee shirts and shorts, some were wrapped in sheets, and some were dressed and wrapped. One pig was neither dressed nor wrapped. Although this study took place in South Africa, some general observations were made that likely relate to other locations.

In the summer trial, maggots were seen by the morning the first day after the pigs were placed. All the summer sample groups showed similar numbers of eggs and flies for the first 13 days. The autumn pigs had more flies per carcass when compared to the summer pigs. Interestingly, all the flies were seen to push their way through small openings to get under the sheets. Bloat lasted less than one day in the summer pigs and about three days in the autumn pigs. There were no really obvious maggot masses in the summer pigs; the maggots were spread fairly evenly over the carcass. There were several differences seen between flies and maggots on the dressed pigs, the greatest being mass loss (decrease in the tissue or weight of the body). Although mass loss was faster in exposed pigs, wrapped pigs lost more overall mass probably due to increased putrefaction and bloating. Lots of other interesting facts were related; check out the article for more information.

There are chapters written about the breakdown of clothing and materials associated with buried bodies; generally synthetic materials take much longer to degrade than natural fibers, like wool and cotton (Janaway, 2008). A body wrapped in clothing can decompose slower due to the lack of contact with soil microbes. Those conditions would likely induce a higher moisture/low oxygen scenario, one associated with the formation of adipocere.

Scavengers

According to one author, scavengers can consume 35–75% of carcasses in land ecosystems and over 100% of carcasses if environmental microbes are not available in the winter (Fitzpatrick, 2008). Scientists have studied the effect scavengers can have on human remains, mainly as a way to understand signs that may be seen on a body (Haglund, 1997; Rathbun and Rathbun,

Figure 3.69 Gnaw marks on ends of bones likely due to rodents or other scavengers. (Photo taken by the author at University of Tennessee's Forensic Anthropology Center).

1997). Obviously, the type of scavengers varies depending on the location of the decedent (Galloway, 1997).

Rodents and other little creatures will generally gnaw on drying bone (Figure 3.69) and can carry disarticulated bones back to nests and dens (Pope, 2010). Carnivores (e.g., fox, coyote, etc.) may "interact" with a body much earlier in the decomposition process by pulling apart and ingesting soft tissue (Galloway, 1997). They will often chew on the face, hands, and feet first. As the body breaks down, scavengers can spread parts of a body over a large area (Pope, 2010). Although most people think of vultures as the main avian scavengers (Figure 3.70), much smaller birds have been known to take hair, clothing, etc., to use as nesting material. Large canids (wolves, coyotes, and fox) are often thought of as doing this, but bears, pigs, and household pets also can have a dramatic effect on body dismemberment and bone scatter (Galloway, 1997).

In water, there are a number of aquatic species that may scavenge a submerged body (Clark, Worrell, and Pleuss, 1997). In salt water (Rathbun and Rathbun, 1997), most people first think of sharks, but in fresh water, smaller predator fish, crustaceans (crayfish, etc.), and turtles can do a lot of damage.

A study was done to look at the scatter and scavenging of the remains of small pigs in a wooded area in the mid-Atlantic region of the United States (Morton and Lord, 2006). Most of the carcasses were placed out in the spring.

Figure 3.70 Scavengers, such as birds, rodents, omnivores, and carnivores, can cause scattering of human remains. Smoky the toy vulture poses (photo courtesy of A.S. Rosenbaum).

The authors reported that once invertebrates (i.e., arthropods or insects, etc.) colonized on the bodies, scavengers pretty much left them alone. Once the bugs had "devoured most of the soft tissue," the scavengers would come in. If there was no colonization of arthropods (too cold), scavengers played a large part in breaking down the body. Bird scavengers (especially crows) were seen eating fly maggots and not remains. On the other hand, groups of vultures collected (often in large numbers) and effectively reduced a body to bones within a matter of days (Figure 3.70). At night, the carcasses were visited by foxes, raccoons, opossums, and skunks. These animals caused more extensive scattering of body parts.

Extreme Conditions

Desert

The desert is most commonly associated with hot, dry conditions, but depending on the location, there can be periods of cold temperatures and rain (Pope, 2010). Factors that affect decomposition in moister climates are similar to those in the desert; temperature and humidity are "inextricably linked" and must be considered together (Galloway, 1997). Elevation (feet above sea level) also plays a role in affecting decomposition (Figure 3.71).

Due to the dryness of an arid environment, the body can dry out quickly. But, according to Galloway, a body on the desert floor may have a rapid period of bloat between two and five days. Then, the top surfaces may begin to dry out and mummify, starting about 11 to 30 days after the body was

Figure 3.71 Low humidity is often found at high altitudes.

placed out in those conditions. As covered in Chapter 2, there may be partial or complete mummification because of decreased gas production and insect activity. Skin starts to disappear over the next several months. Bones can fall apart as the cartilage, ligaments, and tendons holding them together disappear, followed by bones drying out and bleaching. As elsewhere, much of the body breakdown can be aided by carnivore and scavenger activity.

Because rainfall is usually infrequent, but intense, when it occurs in the desert, a body can be carried off by water run-off (Galloway, 1997). This can cause a decomposed body to be scattered over a large area. At higher elevations, melting snowfall can cause a similar effect. In rainy seasons, a microenvironment can be created, especially in shady conditions, creating a moister climate than seen in the surrounding desert (Pope, 2010). In normal desert conditions, the presence or absence of shade can make a huge difference in decomposition rate and how it progresses. In sunny, arid conditions, a body will often mummify and the leathery skin can act as a barrier to insects, further reducing or slowing the removal of soft tissue.

There are not as many differences in decomposition in burials in temperate and arid environments. Because of the 60–80% concentration of water in the human body, it often undergoes moist decomposition with the formation of adipocere, skin slippage, and occasionally fungal growth (Galloway, 1997).

Over time, there is soft tissue lost and the "grease" in bones dries out, but the bones do not usually bleach out. There can be a lot of variation in decomposition seen in a body that is buried and then partially exposed to the surface.

Inside buildings in the desert, bodies may start to bloat in a few days following death and continue for several more days (Galloway, 1997). It then can progress to partial skeletonization pretty quickly after the purge fluid is released; the skin dries out and mummifies. Skeletonization usually occurs first in the head, then the chest and abdomen (Pope, 2010). The pelvic region is usually the last to reach this state of decay. According to Galloway (1997), this could take up to two weeks longer if the body was located outside in arid conditions.

Extreme Heat (Fire)

When a body is exposed to extreme, intense heat, several things can occur (Figure 3.72). One of the most frequently described is when a body is initially exposed to heat and develops the "pugilistic pose;" a condition when the muscles of the arms end up contracting and pulling in toward the chest, supposedly reminiscent of a fighter in the boxing ring (Schmidt and Symes, 2008). If exposure to fire continues, soft tissue is lost, leaving only bone. The process of burning or cremation is more easily understood by examining Table 3.5 (Correia 1997).

When bone was heated during incineration (cremation) to between 300–572°F (150–300°C), there was degradation of the organic matrix (Harbeck et al., 2011). They were able to recover DNA from bone incinerated up to 572°F (300°C).

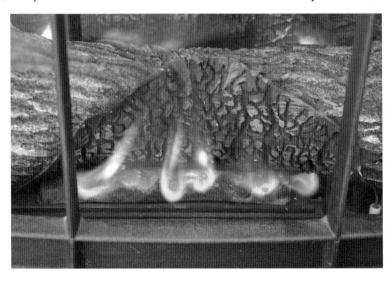

Figure 3.72 Fire can cause many changes to a body. (Photo courtesy of Andrew Rosenbaum.)

Table 3.5   **Stages of cremation and tissue outcome**

Charred	Internal organs
Partial	Soft tissue
Incomplete	Bone pieces
Complete	Ash only

Bone color and consistency changes over time (Correia, 1997). Between 572–1112°F (300–600°C), bone color changed from brown to black to grey to white (Harbeck et al., 2011). Correia reported bone colors changed over time from brown to blue-grey, black to occasionally chalky white. If bone is exposed to intense heat, the bone continues to dry out to a state known as calcination "where the china-like texture of the bone represents a complete loss of the organic portion and the fusion of bone salts" (Correia, 1997).

Freeze and Thaw

At high elevations in very cold climates, bodies are more often preserved by desiccation than freezing (Micozzi, 1997). As the temperature drops below 32°F (0°C), no drying is required because the body freezes instead.

If cold enough, the body can completely freeze and then thaw as the weather warms. During winter, humidity is often pretty low, potentially resulting in partial mummification of exposed skin. Below 55°F (12°C) bacterial growth really slows down; between 32 and 41°F (0–5°C) bacterial multiplication actually stops (Micozzi, 1997). In "normal" (not freezing or thawing) conditions, the intestinal bacteria cause much of the decomposition to occur, kind of decomposing from the inside out. A body that freezes completely is a body that decomposes from the outside in. Intestinal microbes are no longer available to jump start the putrefaction process, so as the soil microbes begin to thaw out, they start the decomposition process (skin inwards). As far as big bugs are concerned, beetles are usually the main arthropod seen feeding on the skin and crawling into cavernous places in the body, helping with decomposition from the outside.

Thanks to Hank the hog (see Table 3.1), the effect of freezing and thawing has already been discussed. When conditions are not as extreme, the intestinal microbes may actually be insulated inside the body. A low level of intestinal microbes may be able to survive until the weather warms up. When this happens, putrefaction can begin and the body can be described as decomposing from the inside out.

Micozzi (1997) described a study in which a freshly dead pig was placed outside along with one that had been frozen. Side-by-side comparisons showed similar changes as seen with Hank and Horace.

Heaving Bones

In bodies buried in climates with a freeze–thaw cycle, the movement of buried disarticulated bones can sometimes be seen over time. Soil freezes first in the top layers and then moves downward. The process can reverse as the weather warms. Disarticulated bones may be pushed up or pulled down by heaving soil, causing mass displacement of the skeleton. According to one source, the bones may concentrate in layers at various depths and move along a "hydro" plane. Bones can thrust to the surface (Figure 3.73 and Figure 3.74), just as rocks pop up in farm fields each spring.

Figure 3.73 Bones from buried bodies can appear at different times and at some distance from the original burial site, due to the freeze–thaw cycle.

Figure 3.74 A close view of bones.

Water

Drowning may be considered a type of burial. As with other topics covered in this chapter, there are books that may be consulted for more information on the drowning process (Armstrong and Erskine, 2011; Teather, 1983; Linton, Rust, and Gilliam, 1986). For this part, only general information is covered, hopefully providing a useful background.

Process of Drowning

Drowning is defined as any death caused by submersion in the water (Linton, Rust, and Gilliam, 1986; Armstrong, 2011). The basic stages are immersion in water, followed by shock, followed by swim failure, then immersion hypothermia, and, finally, postimmersion collapse. Even if the water seems relatively warm, the same process can occur once the person goes underwater.

Submersion is sometimes followed by panic, with the victim responding to water in generally two possible ways. About 70–85% of drownings are "wet drowning where water fills the lungs and interferes with the ability to breathe" (Linton, Rust, and Gilliam, 1986). The other 15–30% are dry drownings in which the "first gulp of cold water will cause a spasm of the larynx, which subsequently suffocates the victim without flooding the lungs" (Linton, Rust, and Gilliam, 1986). When water gets in the lungs, it can mix with the lung air plus a surfactant that lines the inside of the lungs. This surfactant is like a detergent that can serve as a protective layer. When the surfactant mixes with water, it can form a foam or froth that moves up into the upper airways. The spasm of the laryngeal muscles can be followed by formation of a mucus plug, foam, and froth (Armstrong and Erskine, 2011).

Fresh water can alter or destroy this protective surfactant, messing with "alveolar tension," and cause pulmonary edema or an accumulation of fluid in the lungs (Erskine, 2011). The lack of oxygen causes cells in the lungs to switch from an energy system based on oxygen (aerobic) to one that works without oxygen (anaerobic) and producing acid. This acidity (and water) can move into the blood system and increase carbon dioxide (CO_2) concentrations in the blood. These high CO_2 levels can first impair mental function, and then cause a body-wide acidosis. The same things can occur in drowning in sea (salt or saline) water, but usually less extensive injuries are seen (Erskine, 2011).

Generally the body sinks to the bottom and then may go through a similar progression of putrefaction and bloating. Gases produced by internal microbes can become trapped in the lungs and abdominal cavities, causing the body to slowly refloat, possibly reaching the surface of the water

(Erskine and Armstrong, 2011). Rigor, livor, and algor mortis also can occur in drowning victims.

Refloat

There are several factors that can affect whether the body will refloat. Temperature has the greatest effect if a body floats or not. Deeper water is usually cooler than shallow water. In water over 38°F, bacteria in the body should begin to produce gas. Gases diffuse more rapidly and leave the body easier in cooler water than in warm water, "a factor which may also inhibit refloat because of the limited gas buildup" in cool water (Linton, Rust, and Gilliam, 1986). The decedent's last meal also can affect gas buildup; a high carbohydrate diet (as in beans or beer) can provide the bacteria with lots of food to produce gas (Erskine and Armstrong, 2011). The physical condition of the decedent also plays a role in potential refloat time. A high percentage of body fat could insulate and keep the body warmer, helping promote bacterial growth. The health of the victim could affect decomposition as well. If the victim had a fever or internal infection, the elevated body temperature could support bacterial activity and decomposition (Erskine and Armstrong, 2011).

The basic "float" position is with the back and buttocks up, and the face and limbs hanging downward. As the body surfaces, it becomes attractive to insects. If in direct constant light, exposed surfaces may dry out and mummify. The skin will break down through to the deeper layers of the body, creating "holes" for the gas to leak out. As gas is eventually released, the body will usually sink.

Over time, the body may disarticulate or fall apart, usually first the hands at the wrists, head, and lower legs. Clothes can help hold the body together, but may create an environment where adipocere may form. Adipocere can form in warm water in two to three months, while in cooler water, it may take up to 12 to 18 months (Pope, 2010).

Other things that can affect whether the body refloats would include the type of clothing the victim was wearing, if there is a current or tide, how the victim entered the water (jumped in, fell in, etc.), and if there was anything in which the body could become trapped. In a river drowning, the victim may initially sink and then as decomposition occurs and gases become entrapped in the chest and abdominal cavity, the body may start to rise up the water column, and be moved by the currents.

Location in Water

The distance a body can travel also depends on a number of things. One source indicated that a moderate river current can transport a body over a mile in the first half hour (Teather, 1983). The body also can be moved by lake currents or oceanic tides. A body can get trapped in a lake that has holes or deep areas. Water depths around 200 feet may keep a body from floating due to cold temperatures and pressure causing compression of body tissues (Erskine and Armstrong, 2011).

It is also important to understand that a body may not refloat. Any type of penetrating injury that would allow gas to escape from the body could keep it from floating.

The possible location of a drowning victim has been described in a number of books. There are some rules of thumb that could apply to certain drownings, especially those in a lake with a relatively flat bottom (Erskine, 2011). In a lake, the victim is usually found "within a radius equal to the depth of the water level beneath the point where he/she went down" (Teather, 1983). In other words, if a person fell out of a boat in water 30 feet deep, there is a good chance he/she will be located in an area 60 feet in diameter (30-foot radius) under the boat.

Armstrong and Erskine (2011) describe that a body will sink about 2 fps in freshwater and 1.5 fps in salt water (more buoyant). The speed of the current and the "sink rate" may be used to calculate where the body could end up.

The movement of human remains in a river system was further described by Nawrocki et al. (1997). The movement of bodies in large bodies of water is beyond the scope of this chapter.

Appearance

The appearance of the victim once removed from the water can depend on a number of factors (Armstrong and Erskine, 2011). One of those is when a victim gets trapped in a "low head dam." This is a situation when the victim gets trapped under the water surface such as under a waterfall. This results in a water-recycling hydraulic that can keep a victim from floating, and holding the body down, often against an abrasive surface (Erskine, 2011). This action can sometimes cause a sand-blasting effect on the victim, creating abrasions and perforation of the skin. These skin breaks can interfere with gas retention and the body may not refloat.

It is not unusual to find a victim without clothing; it can be totally or partially removed in a number of ways. If a victim has a lot of body fat, decomposition can be faster and result in bloating, This increase in size can keep clothes on the victim. The same can happen in warm water. If the victim jumped into the water, the clothes may be ripped off from the force of entering the water. Current also can pull off clothing. In cases where putrefaction is advanced, river current can "peel off large sections of loose skin" as well (Teather, 1983).

Conclusions

As more research is done, there hopefully will be a better understanding of how the environment affects the taphonomic process. But, as reported in a study of the decomposition of a decedent in central Texas, there were still some "unexpected events" that occurred (Parks, 2011). It seems that taphonomy is not yet a precise, highly predictable science.

References

Alexander, M.1977. *Introduction to soil microbiology.* 2nd ed., New York: John Wiley & Sons.

Anderson, G. S. 2011. Comparison of decomposition rates and faunal colonization of carrion in indoor and outdoor environments. *Journal of Forensic Science* 56 (1): 136–142.

Armstrong, E. J. 2011. Introduction. In *Water-related death investigation: Practical methods and forensic applications,* ed. E. J. Armstrong and K. L. Erskine. Boca Raton, FL: CRC Press, Chap. 1.

Armstrong, E. J., and K. L. Erskine. 2011. *Water-related death investigation: Practical methods and forensic applications.* Boca Raton, FL: CRC Press.

Brookes, P. C. 2008. Principles and methodologies of measuring microbiological activity and biomass in soil. In *Soil analysis in forensic taphonomy: Chemical and biological effect of buried human remains,* ed. M. Tibbett and D. O. Carter. Boca Raton, FL: CRC Press, Chap. 10.

Carter, D. O., D. Yellowlees, and M. Tibbett. 2010. Moisture can be the dominant environmental parameter governing cadaver decomposition in soil. *Forensic Science International* 200: 60–66.

Catts, E. P., and N. H. Haskell. 1990. *Entomology and death: A procedural guide.* Clemson, NC: Joyce's Print Shop, Inc.

Clark, M. A., M. B. Worrell, and J. E. Pleuss. 1997. Postmortem changes in soft tissues. In *Forensic taphonomy: The postmortem fate of human remains,* ed. W. D. Haglund and M. H. Sorg. Boca Raton, FL: CRC Press, Chap. 9.

Correia, P. M. M. 1997. Fire modification of bone: A review of the literature. In *Forensic taphonomy: The postmortem fate of human remains,* ed. W. D. Haglund and M. H. Sorg. Boca Raton, FL: CRC Press, Chap. 18.

Erskine, K. L. 2011. Investigative duties on scene. In *Water-related death investigation: Practical methods and forensic applications,* ed. E. J. Armstrong and K. L. Erskine. Boca Raton, FL: CRC Press, Chap. 2.

Erskine, K. L. and E. J. Armstrong. 2011. On scene body assessment. In *Water-related death investigation: Practical methods and forensic applications.* eds. K. L. Erskine and E. J. Armstrong. Boca Raton, FL: CRC Press, Chap. 3.

Fitzpatrick, R. W. 2008. Nature, distribution, and origin of soil materials in forensic comparison of soils. In *Soil analysis in forensic taphonomy: Chemical and biological effect of buried human remains,* ed. M. Tibbett and D. O. Carter. Boca Raton, FL: CRC Press, Chap. 1.

Galloway, A. 1997. The process of decomposition: A model from the Arizona-Sonoran desert. In *Forensic taphonomy: The postmortem fate of human remains,* ed. W. D. Haglund and M. H. Sorg. Boca Raton, FL: CRC Press, Chap. 8.

Gill-King, H. 1997. Chemical and ultrastructural aspects of decomposition. In *Forensic taphonomy: The postmortem fate of human remains,* ed. W. D. Haglund and M. H. Sorg. Boca Raton, FL: CRC Press, Chap. 6.

Hadley, D. n.d. How are accumulated degree days calculated? Online at: http://insects.about.com/od/forensicentomology/f/ADDcalculation.html (accessed on June 6, 2011).

Haglund, W., ed. 1997. Dogs and coyotes: Postmortem involvement with human remains; Scattered skeletal human remains: Search strategy considerations of locating missing teeth; Rodents and human remains. In *Forensic taphonomy: The postmortem fate of human remains.* Boca Raton, FL: CRC Press, Chaps. 23, 24, 26.

Harbeck, M., R. Schleuder, J. Schneider, I. Wiechmann, W. W. Schmahl, and G. Grupe. 2011. Research potential and limitations of trace analyses of cremated remains. *Forensic Science International* 204: 191–200.

Haskell, N. H., R. D. Hall, V. J. Cervanka, and M. A. Clark. 1997. On the body: Insects' life stage, presence, and their postmortem artifacts. In *Forensic taphonomy: The postmortem fate of human remains,* ed. W. D. Haglund and M. H. Sorg. Boca Raton, FL: CRC Press, Chap. 27.

Janaway, R. C. 2008. Decomposition of materials associated with buried cadavers. In *Soil analysis in forensic taphonomy: Chemical and biological effect of buried human remains,* ed. M. Tibbett and D. O. Carter. Boca Raton, FL: CRC Press, Chap. 7.

Judah, J. C. 2008. *Buzzards and butterflies: Human remains detection dogs.* Brunswick County, NC: Coastal Books.

Kelly, J. A., T. C. VanderLinde, and G. S. Anderson. 2009. The influence of clothing and wrapping on carcass decomposition and arthropod succession during the warmer seasons in central South Africa. *Journal of Forensic Science* 54 (5): 1105–1112.

Linton, S. J., D.A. Rust, and T. D. Gilliam. 1986. *Dive rescue specialist: Training manual.* Fort Collins, CO: Concept Systems, Inc.

Megyesi, M. S., S. P. Nawrocki, and J. H. Haskell. 2005. Using accumulated degree-days to estimate the postmortem interval from decomposed human remains. *Journal of Forensic Science* 50 (3): 618–626.

Metting, F. B. 1993. Structure and physiological ecology of soil microbial communities. In *Soil microbial ecology: Application in agricultural and environmental management.* New York: Marcel Dekker, Chap. 1

Miccozzi, M. S. 1997. Frozen environments and soft tissue preservation. In *Forensic taphonomy: The postmortem fate of human remains,* ed. W. D. Haglund and M. H. Sorg. Boca Raton, FL: CRC Press, Chap. 11.

Michaud, J. P., and G. Moreau. 2011. A statistical approach based on accumulated degree-days to predict decomposition-related processes in forensic studies. *Journal of Forensic Science* 56 (1): 229–232.

Morton, R. J., and W. D Lord. 2006, Taphonomy of child-sized remains: A study of scattering and scavenging in Virginia, USA. *Journal of Forensic Science* 51 (3): 475–479.

Nawrocki, S. P., J. E. Pleuss, D. A. Hawley, and S. A. Wagner. 1997. Fluvial transport of human crania. In *Forensic taphonomy: The postmortem fate of human remains,* ed. W. D. Haglund and M. H. Sorg. Boca Raton, FL: CRC Press, Chap. 34.

Parks, C. L. 2011. A study of the human decomposition sequence in central Texas. *Journal of Forensic Science* 56 (1): 19–22.

Pope, M. 2010. Differential decomposition patterns of human remains in variable environments of the Midwest. Orlando: University of Southern Florida. Online at: from http://scholarcommons.usf.edu/etd/1741 (accessed June 11, 2011).

Rathbun, T. A., and B. C. Rathbun 1997. Human remains recovered from a shark's stomach in South Carolina. In *Forensic taphonomy: The postmortem fate of human remains,* ed. W. D. Haglund and M. H. Sorg. Boca Raton, FL: CRC Press, Chap. 28.

Rodriguez, W. C. 1997. Decomposition of buried and submerged bodies. In *Forensic taphonomy: The postmortem fate of human remains,* ed. W. D. Haglund and M. H. Sorg. Boca Raton, FL: CRC Press, Chap. 29.

Schmidt, C. W., and S. Symes. 2008. *The analysis of burned human remains.* Burlington, MA: Academic Press.

Simmons, T., R. E. Allison, and C. Moffatt. 2010. Debugging decomposition data— comparative taphonomic studies and the influence of insects and carcass size on decomposition rate. *Journal of Forensic Sciences* 55(1):8–13.

Teather, R. G. 1983. *The underwater investigator.* Fort Collins, CO: Concept Systems, Inc.

Tibbett, M., & D. O. Carter. 2008. *Soil analysis in forensic taphonomy: Chemical and biological effect of buried human remains.* Boca Raton, FL: CRC Press.

Vass, A. A., W. M. Bass, J. D. Wolt, J. E. Foss and J. T. Ammons. 1992. Time since death determinations of human cadavers using soil solution. *Journal of Forensic Science* 37(5): 1236–1253.

White, T. D., and P. A. Folkens. 2005. *The human bone manual.* New York: Elsevier Academic Press.

Wilson, A. S., R. C. Janaway, A. D. Holland, H. I. Dodson, E. Baran, A. M. Pollard, and D. J. Tobin. 2007. Modelling the buried human body environment in upland climes using three contrasting field sites. *Forensic Science International* 169: 6–18.

Making Sense Out of Scent

4

The A, B, C, Ds

A: About

This chapter is **A**bout the role of **B**acteria and the **C**hemicals that may be produced during the **D**ecomposition process. It also includes a review of some of the research that has been done to characterize the chemical profile of odor or scent human remains (HR).

Note that:

1. The information in this chapter is a "generalized description" of what has been found in some scientific studies.
2. To the author's knowledge, there is no consensus among the scientific community as to what makes up the chemical profile for human remains.
3. It is assumed that the chemicals identified in off-gassing from HR makes up (in part?) some of the chemical signature.
4. The chemical signature may be a result of the breakdown of tissues and/or microbial by-products.
5. It is assumed that the chemicals can be altered by environmental conditions.
6. It is not known what human remains detection (HRD) dogs are actually keying in on.

The intention of this chapter is to help handlers and investigators have an understanding of what may possibly make up the chemical signature or odor profile of decomposing human remains. The research in this area is very new and there are certainly more questions than answers. The information in this chapter is *not* the gospel of how the dog actually detects and locates human remains. Simplify, this is a very basic review of what *may* make up some of the chemicals that become available throughout the decomposition process. With this information, the reader should have a better understanding and be able to describe the complexity of conditions that the HRD dog works in as a locating tool.

B: Bacteria and Bodies

Decomposition is the process of big things getting smaller. As we already know, decomposition of a body is dependent on a number of factors, and one of the most important is bugs. Without big bugs (insects, etc.) or little bugs (bacteria, etc.), the natural decomposition process could take much longer and be much different from what naturally happens. The role of insects on decomposition was covered in Chapter 3; the little bugs are the focus of this chapter.

Little Bugs

As we know, soft tissue (including pretty much everything except bone) is often broken down by bacteria, fungi, and protozoa—microscopic organisms or microbes found in the body and the environment. Microbes are found almost everywhere and play a useful role in maintaining balance in our lives (Figure 4.1). There are literally thousands of types of bacteria in nature and

Figure 4.1 Flora and fauna (including microbes) vary tremendously throughout the environment.

many types prefer to live in an environment that is moist and warm. The microbial population of interest are those in the body itself and in the environment (especially soil).

Microbes can include bacteria, protozoa, yeast, etc., but will be generically referred to as bacteria and microbes. Some prefer to live in conditions where there are high levels of oxygen (aerobic microbes), while others prefer to live where there are low levels of oxygen (anerobic microbes). If the oxygen levels change, there can be a shift in the type of microbes found in a specific area. Facultative microbes are ones that can live in anerobic or low O_2 (oxygen gas) conditions, but then can go crazy reproducing when given O_2. Changes in O_2 levels are a common occurrence during decomposition because O_2 levels decrease throughout the process, especially in the bacterial populations in the gastrointestinal (GI) and respiratory systems. If there is a change in temperature, microbial populations can change as well. High temperatures can be as destructive to some strains of bacteria as cool temperatures (Metting, 1993).

There are good bacteria that help animals digest certain foods, e.g., cattle. Cows are dependent on a vast microbial population in their GI tract that actually digests the hay and corn that they eat (Figure 4.2). Then as the bugs do their thing, the microbes produce nutrients that the cow needs to live. An advantageous arrangement: Feed the bugs and the cows feed us.

To understand and explain the role of microbes, let's go back to the cow. The cow eats grass and hay, something that we can't do and expect to survive. How does she do it? Well, it is because the cow has four compartments in her

Figure 4.2 Cows are able to digest plants that people can't due to the huge number and variety of microbes in their stomachs.

stomach; each has a different job, but the main one is to make the pieces of grass get smaller and smaller. As a cow eats the grass, she chews it, swallows it, regurgitates it as cud, rechews and then reswallows it, until the grass particles get smaller and smaller. The pieces finally get small enough for the microbes in the rumen (one of the stomach compartments) to work on the small bits of grass.

A little known fact is that the rumen in an adult Holstein dairy cow is about the size of a 55-gallon drum (Figure 4.3).That drum is essentially a fermentation vat where the microbes can eat up and break down small pieces of grass. As an end result of digestion, the bacteria produce gas (lots of it), and break down products into carbohydrates, proteins, and fats. These products can be absorbed and used by the cow, something she couldn't do without those little bugs.

Production of breakdown products in cows' rumens has been studied for alternative energy (Thompson, 2007). Microbial fuel cells with 16 ounces of ruminal contents (and millions of microbes) produced about 600 millivolts of electricity, about half the voltage of a rechargeable AA battery (Owens, 2005).

Maybe another explanation will help. Instead of cows, how about beer (Figure 4.4)? Making beer uses a similar process: The microbes in the fermentation tank are what breaks down the carbohydrates (from corn, hops, etc.) and produce gas and alcohol (Brisson and Nibbe, 2003). This is basically the same role that bacteria play during the decomposition of a body in helping to break larger compounds into smaller compounds.

Yeast is responsible for "converting fermentable sugars into alcohol and carbon dioxide" (Brisson and Nibbe, 2003). There are two major types of

Figure 4.3 About the size of a 55-gallon drum, the rumen works as a huge muscular fermentation vat.

Figure 4.4 Making beer uses a similar process to bacteria's role in breaking large compounds into smaller ones.

Table 4.1 Role of yeast in production of ales

Type of Yeast	Fermentation	Temperature	Comments
Ale	Top	10–25°C	Rich, thick, bready smell with lots of foam, ales, stouts, porters
Lager	Bottom	7–15°C	Works slowly, little foam, malt liquors, lager, pilsner

yeast that, when allowed to work at certain temperatures, produces two very different products. Table 4.1 shows the end results. The food for these yeasts is commonly corn. Interesting, as it is similar to the cow. Composting of organic matter (ranging from kitchen waste to bodies) also depends on microbes that break things down also through fermentation (Miller, 1993).

This section provides probably an over-simplistic explanation, but the goal is to provide the basics to help one understand where the chemicals that may make up the odor profile or chemical signature of human remains may come from. Before going any farther, it is time to step to the **C** part of this chapter.

C: Chemistry (Made Simple)

Elements, molecules, compounds, families, classes, volatiles: What does it all mean and why should you care? Because it is a necessary part of understanding decomposition.

Definitions

Organic is a term often heard. It basically means that a chemical molecule or compound has a carbon (C) atom. A hydrocarbon is a molecule with C and hydrogen (H). Other elements of focus in this chapter will be limited to a few of the numerous elements in the periodic table: C and H as well as nitrogen (N) and sulfur (S). How chemists have determined what chemical structures look like and how they act is a science beyond the scope of this book. However, there is some basic information that will help the reader understand the results of scientific studies that identified some of the specific chemicals that have been associated with human remains.

Carbon is the "element of life" and fairly easy to understand. The C atom has four electrons in its outer shell; these electrons can be thought of as parking spaces with the C at the center. If a C bonds with four H atoms, you get the hydrocarbon gas methane (CH_4).

C also can bond to other C to form long chains. The backbone of any organic compound is the carbon–carbon bond (C–C), that varies by the number of C atoms in the sequence (Bloch, 2006). The C atoms can be connected together either by a single, double, or triple bond (Figure 4.5). The type of bond can determine its strength, basically how long a compound will hold together, and what it may take to pull it apart. This will become important in Chapter 5.

If C–C bonds is single, it can be designated as an alkane. With a single C–C, the backbone of the molecule is straight; straight chains are called

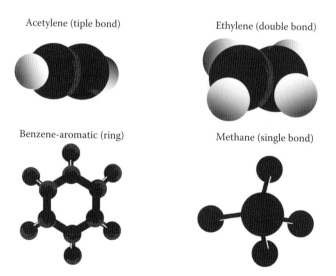

Acetylene (tiple bond)   Ethylene (double bond)

Benzene-aromatic (ring)   Methane (single bond)

Figure 4.5 Carbon: The chemical of life. This shows the difference bonds can make in the chemical. (From structures modified from various online sources.)

Table 4.2   Numbers of carbons and the functional groups

Number of Carbons	Functional Group	Chemical Formula
1	Methane	CH_4
2	Ethane	C_2H_6
3	Propane	C_3H_8
4	Butane	C_4H_{10}
5	Pentane	C_5H_{12}
6	Hexane	C_6H_{14}
7	Heptane	C_7H_{16}
8	Octane	C_8H_{18}
9	Nonane	C_9H_{20}
10	Decane	$C_{10}H_{20}$

aliphatic compounds. Another important thing about the C–C single bond molecule is the concept of saturation, i.e., when all the available parking spaces are taken up by hydrogen atoms, it is considered saturated.

The C=C double bond (alkene) and C≡C triple bond (alkyne) are not saturated because there is less room on the molecule for H atoms. When this happens, instead of a line or chain, the C backbone becomes a ring or circular structure (Figure 4.5).

The number of C atoms in a compound can be used to name or describe what a chemical looks like and how it may act. Table 4.2 includes the prefix that designates the number of C atoms. Some of this may now begin to look familiar.

Functional Groups

Along the C backbone, there may be arms and legs that hang from it called functional groups. Whether straight or connected (1×, 2×, or 3× bonds), these functional groups also help define what the molecule is and how it may act in the environment.

The easiest functional group is the hydroxyl group—oxygen and hydrogen (-OH) combined together to bond to a carbon. The hydroxyl group is more commonly called an alcohol. Therefore, using the number of Cs in the molecule, we can name it: CH_3-OH = methyl alcohol. CH_3-CH_2-OH would be ethyl alcohol or ethanol. Hydroxyl groups are pretty easily removed when exposed to water.

The carbonyl group is simply C and O, and a carboxyl group is COOH. The carboxyl group is characteristic of "all organic acids" (Carpi, n.d.). A methyl group (CH_3) hanging off a benzene (C_6H_6) ring is methylbenzene, also known as toluene.

A Family Affair

This section will cover some of the basics of various chemical families found in organic chemistry. Many of these families have been associated with human remains. They include:

- **Alkanes:** These are molecules that contain only H and C and can eventually break down to CO_2 and H_2O (Carpi, n.d.). Many of the common alkanes are flammable and commonly used as fuels. Some examples include methane (CH_4, Figure 4.5), butane (C_2H_6), and propane (C_3H_8).
- **Alkenes:** These are hydrocarbons with one or more double bonds between carbons *and* that are a straight chain. Alkenes can be saturated or unsaturated; because of this, these compounds readily undergo chemical reactions in water or alcohol. The simplest is ethylene ($H_2C=CH_2$, Figure 4.5), which is created when crude oil is heated and degrades.
- **Alkynes:** These are compounds that have a triple bond between carbons. The most commonly recognized example is acetylene (Figure 4.5).
- **Acid/Acid Esters:** Example of organic acids is acetic acid or vinegar. Esters are likely present as part of saponification (Statheropoulos, Spiliopoulou, and Agapiou, 2005).
- **Alcohols:** As previously mentioned, alcohols have a hydroxyl (OH) group. Ethanol (CH_3CH_2OH) is a fermentation product of grains.
- **Aldehydes:** These are highly reactive organic compounds that are created by the addition of an O atom to certain alcohols to form a CHO group. A common example is formaldehyde.
- **Amides:** These are organic compounds that contain N, O, and H: a $C-ONH_2$ radical group. Amides are often volatile solids. A common example is acetamide, a crystalline amide of acetic acid (CH_3CONH_2).
- **Aromatic compounds:** Aromatic compounds have alternating single and double bonds between C atoms (Anon., 2006). The basic six-sided cyclo-structure is known as a benzene ring (Figure 4.5). These are unsaturated cyclic compounds that are more stable than straight chain alkenes. The aromatic part of the name comes from the "sweet" odor that some of these compounds have.
- **Ester:** An ester is formed when an acid and alcohol combine and release water. Generally these are a short-chain compound associated with pleasant aromas and food flavorings. Large esters can form from long chain carboxylic acids and commonly occur as animal fats.
- **Ketones:** This is an organic compound with a carbonyl (CO) group linked to a C atom. A common example is acetone (CH_3COCH_3).

- **Radical:** An atom or group of atoms with >1 unpaired electron. Radicals are generally pretty reactive.
- **Sulfides:** The presence of sulfur (S) in a molecule can be indicated by the term *sulf* or *thio*. Some examples include disulfide or sulfonic acid. Sulfides help give the undesirable odor to rotten eggs as well as skunks.
- **Volatile:** This means that something readily evaporates at normal temperature and pressure conditions. A volatile organic compound (VOC) easily vaporizes.

Shape and What It Means

It's important to have a grasp of the above to understand the rest of the chapter: What is it that could be making up the odor of human remains?

Under the best of conditions, any organic compound can theoretically continue to degrade completely to water and carbon dioxide (H_2O and CO_2) only. This process is called mineralization. An example? And an interesting one? Microbes (bacteria and yeasts) in municipal anaerobic sludge have been shown to completely biodegrade the explosive RDS (hexahydro-1,3,5-trinitro-1,3,5-triazine) down to H_2O and CO_2 (Hawari et al., 2000).

Although this will be touched upon in Chapter 5, the structure of a compound (straight or cyclic, saturated or unsaturated) will affect how a compound reacts in the environment. A saturated compound (with all the H available in the molecule) is generally hydrophilic, meaning that compound loves water. A saturated aliphatic compound can be picked up by water, carried through the environment, and is likely to break down or degrade. An unsaturated, cyclic compound is generally described as lipophilic, meaning "fat loving." It may be more resistant to breaking down and may stay in its original form longer. Lipophilic cyclic compounds are often persistent in the environment and do not break down and move like aliphatic compounds.

What chemicals are the ones HRD dogs detect? How do they disperse? Not easy questions to answer.

D: Decomposition

At this point, it is time to understand how to describe the concept of an odor that is made of a mix of many different chemicals? To do that, we are back to food, this time vegetable soup. Depending on who makes the soup and what their preferences are, soup can consist of many different vegetables. What all soup does have, however, is broth. Then there are probably carrots, onions, potatoes, and maybe, if you like them, Brussels sprouts. Remember this example, we will return to it shortly.

Live Scent

Before we get started on chemicals associated with human decomposition, let's touch upon live human scent. As has been known by dogs for centuries and now shown by science, each human has a distinct scent. It appears that there is a genetic component that produces each person's scent, which is as "individual as a fingerprint" (Harvey et al., 2006). This distinct scent profile is produced by a number of primary, secondary, and tertiary factors. Primary factors are those "that come from within and are stable over time regardless of diet and environment" (Curran, Prada, and Furton, 2010). This is the odor of a person. Secondary factors include the diet of the individual and in what environment they live in. Tertiary factors are things that may be added to the skin, such as lotions, soaps, etc. Bacteria also are involved in the live scent profile as bacteria are found on the dead skin cells that we shed (Figure 4.6). Curran, Prada, and Furton described how rafts of dead epithelial skin cells are continuously shed and that the rafts contain four to five times as many "germs" as in the air in the rest of the sampling room. There is little doubt that bacteria play a role in live scent.

In between

The transition between life and death likely is accompanied with a change in the VOCs that off-gas from the body. As we know, flies can appear within

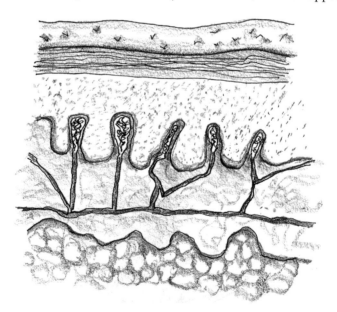

Figure 4.6 Dead skin cells (squamous epithelium) from the top layer of skin can flake off and get carried away.

minutes after death; it is believed that their signal is a change in the odor profile. It is likely that the chemical profile begins to change as the cells move through the reversible to irreversible phases described in Chapter 2.

Not Live Scent

Microbes (including bacteria) have a huge role in the decomposition odor. At this point, it is time to take a reverse engineering approach. How a body breaks down into smaller and smaller parts and eventually may produce chemical compounds that may be the target odor for HRD dogs is the focus of this chapter. As covered in Chapter 2, the body is made up of systems that are made up of organs that are made up of cells that contain carbohydrates, proteins, fats, and other elements. As different types of cells decompose, they probably end up as the chemicals that can or may contribute to the odor of human decomposition.

Note: It is essential to understand that the following are examples of what could be produced during the decomposition process and that different conditions, different tissues, different species, etc. can affect the outcome. The following information is a "best guess" at this time.

Gas

As stated above, when microbes digest and reproduce, they produce gas. Some of the most common gases include carbon dioxide (CO_2), hydrogen sulfide (HS), ammonia (NH_3), methane (CH_4), sulfur dioxide (SO_2), and hydrogen (H) (Statheropoulos, Spiliopoulou, and Agapiou, 2005).

Another concept to understand is how most of this research was in collecting and analyzing air samples. Let's go back to the vegetable soup. Consider that the air coming off any type of remains is that soup. If we were going to figure out how much and what vegetables are in that soup, we would pour the soup through a colander or sieve and get rid of the broth or stock. Then, we would go to the colander and sort out the vegetables by type: carrots in one pile, onions in another, and potatoes in another. Then we would count how many of each type there are: 18 carrot chunks, 15 onion rings, 29 hunks of potato.

Sampling air is similar to counting the vegetables in the soup. The air (soup) is collected from over the tissue or remains and stored until it can be analyzed. Then, during the analysis part, the air (soup) is separated off while the chemicals (vegetables) are trapped. After being trapped, the chemical are separated into types, and quantified or counted. This is a really oversimplified short course in volatile chemistry.

So, researchers have collected air samples from a number of different tissues, kept in a number of different situations, and from tissues from a number of different species. Scientists have figuratively strained off the broth and identified which vegetables were there and how many of each was present. The rest of the chapter includes information from a number of studies that have been done over the past decade. Highlights from some of those studies are included at the end of the chapter.

The approach for this section is to review the major biomolecules in the body, what they are, where they are, and how they may break apart. These include carbohydrates, proteins, and lipids.

Carbohydrates

Carbohydrates (Figure 4.7) are a natural product with many functions in plants and animals and are made up of various configurations of H and C atoms. (Hart, 1983). While the primary carbohydrate in plants is cellulose, the sugar glucose is the one in animals. Called a hydrate of carbon, glucose is fully saturated and has 6 C and 12 H atoms. The molecular formula for glucose is $C_6H_{12}O_6$. Carbohydrate molecules can get pretty complex, but can be classified by the number of carb units. Monosaccharides are single C units, oligosaccharides have two to eight Cs, and polysaccharides have multiple C units.

The body stores glucose to use for long-term energy in the form of glycogen. This molecule can be up to 100,000 glucose units long. Glycogen is commonly stored in the liver and muscle tissue. Glucose units are released as needed (Forbes, 2008).

Figure 4.7 Carbohydrates are commonly associated with plants as in bread and potato pancakes. Protein is prevalent in meat, including corned beef.

Carbohydrate structures can get pretty complex, mainly because of the functional groups attached to the C–C backbone. The chemistry of carbs (how they act and break down) is heavily influenced by the hydroxyl (OH) and carbonyl (OC) groups.

Carbohydrate Breakdown

Degradation of the basic carbohydrate takes place as expected. The polysaccharide glycogen will break down to oligosaccharides to monosaccharides and, if conditions are right, finally to CO_2 and H_2O. When carbohydrates break down in aerobic conditions (with oxygen), glucose is converted to organic acids (citric, oxalic, or glucoronic acids) or alcohols. These may eventually degrade to CO_2 and H_2O (Hart, 1983). Carbohydrates also can break down into aldehydes, ketones, acid esters, and ethers (Statheropoulos, Spiliopoulou, and Agapiou, 2005). In anaerobic conditions, acetic, butyric, and lactic acids can form along with alcohols (ethanol and butanol) and gases (hydrogen, sulfur compounds, and methane [CH_4]) (Forbes, 2008).

Proteins

Protos is Greek for first or of prime importance (Hart, 1983). The name is well deserved as proteins provide many functions in the body. Proteins (Figure 4.7) are complicated biomolecules made up of strings of amino acids that have many different functional groups, each playing a critical role in biochemical reactions in the body.

Amino Acids

Amino acids are simply a molecule that has an amino group (NH_3 – an ammonium group) and a carboxyl group (COO-). Because of this unique combination, amino acids are described as amphoteric. They can act as an acid or a base depending on what type of environment they are in. This simple ability gives proteins the specificity they need to function.

Simple amino acids hooked together are called peptides. As peptides combine, they are called polypeptides. Proteins can combine with other types of biomolecules and structural arrangements to provide a variety of functions for many forms of organic life. Proteins are very diverse and complex structures.

Types of Proteins

There are two basic classes of proteins: fibrous and globular (Hart, 1983). Fibrous proteins are water insoluble and provide structure in the form of keratin in skin, hair, and nails, and collagen in cartilage, tendons, and blood vessels (Figure 4.8). Fibrous proteins are built to be very tough, sometimes reinforced with disulfide crosslinks and helical (screw-like) configurations.

Figure 4.8 Horse hooves (and hair) are forms of fibrous protein.

This framework gives the fibrous proteins some flexibility and strength along with a toughness or resistance to break down easily.

Globular proteins are very different from the fibrous; they are the proteins that tend to be water soluble, globby in shape, and four main biological jobs. Those jobs include working as enzymes (which help speed up and carry out biological reactions), as hormones, as transport proteins, and as storage proteins. Globular proteins are the ones in cells that affect chemical reactions in the body, carry necessary nutrients, etc., in and out of the cell, and store stuff as needed for the future. There is a lot of variation and complexity among globular proteins. One of the more commonly recognized globular proteins is hemoglobin, the O_2-carrying protein found in red blood cells.

Protein Breakdown: Basics

Protein is broken down by chemical enzymes called proteases. There are many different types of proteases, but to keep to the basics, proteins are broken down by proteases to polypeptides to peptones to amino acids (Forbes, 2008). Although proteins can be broken apart in a number of different ways, one of the simplest ways is through hydrolysis. A hydrolytic reaction involve the breakage of the peptide or amide bond between two amino acids, resulting in an amino acid connected to a water molecule. This obviously require that protein to be hydrophilic or water loving. More on this in Chapter 5.

Bacteria (especially *Clostridium* spp. and *Lactobacillus* spp.) can produce enzymes that help break down amino acids (DeGreeff, 2010). Those enzymes can pull off functional groups containing N (deamination), CO (decarboxylation), S (desulfhydration), and, possibly, ammonia, all resulting in a release of something called biogenic amines. Some of these amines break down further to products associated with decomposition. Two amino acids reportedly associated with human remains include skatole and/or indole (Hoffman et al., 2009). Not surprising, sulfur-containing amino acids can form dimethyldisulfide, hydrogen sulfide gas, and sulfuric acids in soil (Statheropoulos, Spiliopoulou, and Agapiou, 2005). In anaerobic conditions, nitrogen-containing amino acids can be broken down to pyruvic acid and ammonia (NH_3). The ammonia can be converted to ammonium (NH_4+) in acidic soil, providing a very desirable fertilizer for plant growth.

Two other by-products of protein breakdown include putrescine and cadaverine. These are "decarboxylated diamines," commonly labeled as VOCs associated with decedents (Forbes, 2008). These are formed when the biogenic amines ornithine and lysine break down.

Protein Breakdown: Tissues

Proteins break down differently in different types of cells in the body. The first proteins to decompose are in nerve and epithelial cells, those most commonly found in the brain, GI system, liver, and kidneys. Then, proteins in muscle, connective tissue, and cartilage break down. Breakdown of muscle has resulted in the production and/or release of acid esters, alcohols, aldehydes, few aromatic hydrocarbons, and ketones (including toluene, and p-xylene) (Hoffman et al., 2009). Hair, nails, and skin are usually some of the last tissues to decompose in the body, all due to the insoluble fibrous protein keratin.

Specific Tissues

Blood

Table 4.3 shows the variability in products found by breakdown of different types or states of human blood (Hoffman et al., 2009). What is shown is the different numbers of chemicals (classes) found.

Table 4.3   Number of compounds found in chemical classes among different "types" of human blood*

Chemical Class	Human Blood	Human Blood Clot	Human Placental Clot
Acid/acid esters	1	—	—
Alcohol	4	—	—
Aldehydes	3	1	—
Aromatic hydrocarbons	2	2	2
Halogens	1	1	1
Ketones	1	—	—
Sulfides	—	1	

* Adapted from Hoffman, 2009.

Take home message: Not all human blood is the same!

Bone

Most people think about bone (Figure 4.9) as being a solid hunk of calcium that provides support for the body. Instead, it is a "living, dynamic tissue" made up of organic and inorganic parts (France, 2010). Calcium and phosphorous make up the inorganic part that forms around an organic matrix made up of (in part) the protein collagen.

Figure 4.9 Called a living dynamic tissue, bone has a hard outer mineral part that encloses the cellular marrow.

In the very young, bones develop from a cartilage frame, growing in length from different growth centers. Different bones mature at different rates, but basically they mature as calcium fills in the collagen framework. Bones also differ in density. The part of bone that is very dense (usually the part of long bones surrounding the marrow cavity) is called cortical bone. Other bones that have kind of a spongy appearance with lots of air spaces are called trabecular or cancellous bone.

The collagen in bone will break down before the hardened mineral parts. As bone goes through the decomposition process, the marrow breaks down first, giving the bone a brownish, greasy appearance. Eventually it dries out.

Hoffman et al. (2009) measured three human bone samples that had certain VOCs in common including aldehydes, alcohol (1), halogenated compound (1), aromatic hydrocarbon (1), and sulfide (1). Vass et al. (2008) found 72 chemical compounds from bone only, but considered these listed in Table 4.4 to be important in differentiating bones from humans and animals.

Probably of most interest to HRD K9 handlers is Table 4.5. In a study, Vass et al. (2008) collected and analyzed the VOCs coming off a number of different types of nonhuman bones. Cablk, Szelagowski, and Sagebiel (2012) also showed that there were differences in the chemical profile of air sampled from cattle, chicken, and pig bone. Again, the numbers aren't the focus, but the differences between the species.

Table 4.4   "Markers" of burial decomposition from human bone[*]

Ketones	Amides	Aldehydes	Alcohols
2-propanone	N,N-dimethyl-acetamide	Hexanal	Phenol
2-decanone	acetamide	Hepantal	Heptanol
2-nonanone		Nonanal	Hexanol
		Octanal	Ethanol
		Pentanal	
		Decanal	
		Butanal	

[*] Adapted from Vass, 2008.

Table 4.5   Percentage of chemical classes among various human and nonhuman bones[*]

Species	Aldehydes	Amines	Alcohol	Ketones
Human	4	0	5	28
Dog	7	46	44	3
Deer	39	23	9	42
Pig	50	31	42	27
Total	100	100	100	100

[*] Adapted from Vass, 2008.

Lipids and Fats

Fat or adipose tissue is made up of really "thin-skinned" cells with basically globs of fat inside (Figure 4.10). Glycerol + fatty acids = fat. Triesters of glycerol or triglycerides make up about 90–99% of lipids and about 60–85% of body fat is made up of lipids (Forbes, 2008). The physical state of the fats will depend on the degree of saturation of H atoms in the fat molecule; generally, saturated fats (with many Hs)will be solid at room temperature.

Table 4.6 (Hart, 1983) includes some of the commonly found fatty acids along with how many C atoms it has as well as its melting point in Farenheit (F) and Celsius (C). Generally, oils are from plants and are liquid at room temperature. Fats are solid and come from an animal source.

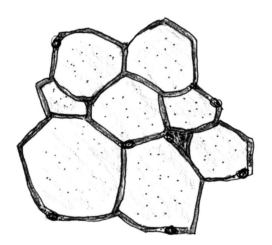

Figure 4.10 Fat cells (adipocytes) have very thin cell membranes, small nuclei, and very large vacuoles containing fat. (Adapted from Hole, J. W. 1990. *Human anatomy and physiology*. Dubuque, IA: Wm. C. Brown Publishing.)

Table 4.6   Commonly found fatty acids with melting points*

Physical State	Fatty Acid	Number of C	Melting Point (°F/°C)
Saturated	Lauric	12	111/44
	Myristic	14	136/58
	Palmitic	16	145/63
	Stearic	18	158/70
	Arachidonic	20	171/77
Unsaturated	Oleic	18	55/13
	Linoleic	18	23/−5
	Linolenic	18	12/−11

* Adapted from Hart, 1983.

Like other biomolecules, lipids can have functional groups attached to them. One of the more common functional groups is phosphatidylamine. As indicated by the term *amine* and the phosphorous part, this is kind of a protein/lipid combination molecule. Phospholipids are a major component of cell membranes and create a bilayer by folding over on themselves. Phospholipids have both hydrophilic and lipophilic parts that each love water or fat.

Fat Breakdown

Lipids are broken apart by enzymes called lipases. Lipases work by hydrolyzing (remember breaking things apart with water) down into fatty acids and glycerol (Hart, 1983). Throughout the degradation process, "breakdown products" can include acid esters, alcohols, aldehydes, aromatic HCS, ketones, and sulfides. VFAs break down to phenol and glycerol (Statheropoulos, 2007). Eventually, these products could mineralize to CO_2 and H_2O.

Fats are broken down by aerobic and anaerobic microbes to hydrocarbons and N and P compounds. Aerobic microbial degradation causes unsaturated fatty acids to oxidize to aldehydes and ketones (DeGreeff, 2010). Unsaturated fatty acids can become saturated in anaerobic conditions. Water also can cause a reaction where chemical bonds in fat can be broken apart; this is called hydrolysis (Jackson, 2001). Hydrocarbons can further break down into volatile fatty acids (VFAs), which can eventually break down into phenol compounds and glycerols.

The fatty acids most commonly associated with human decedents are oleic acid (unsaturated C18:1), linoleic acid (polyunsaturated C18:2), and then lesser amounts of palmitoleic acid (monounsaturated C16:1) and palmitic acid (saturated C16:0) (Forbes, 2008).

Table 4.7 is from a study by Hoffman et al. (2009) and shows the variety of compounds found to off-gas from different forms of human lipids.

As covered in Chapter 3, adipocere formation is also called saponification and can form in low oxygen and high moisture conditions. In 1917, Ruttan and Marshall did some experiments to identify the actual composition of adipocere.

Table 4.7 Compounds found in the following "types" of human fat*

Fat	Fat and Skin (Combined)	Adipocere
Butanoic acid	Pentanoic acid	Butanoic acid
Butanoic acid: ethyl ester	Hexanoic acid: ethyl ester	Butanoic acid: butyl ester
Hexanoic acid		
Butanoic acid: butyl ester	Hexanoic acid: butyl ester	Hexanoic acid: ethyl ester
Hexanoic acid: ethyl ester	Hexanoic acid: hexyl ester	Pentanoic acid

* Adapted from Hoffman, 2009

They found that "hard, clean adipocere wax" was about 68% palmitic acid, 4% calcium salts, and about 25% other fatty acids.

In 2001, Takatori found that adipocere forms from various aerobic and anaerobic bacteria that are able to convert some unsaturated fatty acids to hydroxylated fatty acids (10-hydroxystearic acid and 10-hydroxypalmitic acids). There are a few other types of fatty acids that contribute to the adipocere's hardness that keeps it stable in certain environmental conditions (Ruttan and Marshall, 1917). The melting point of adipocere is approximately 60–63°C.

Summary

What we don't know about the specifics of the chemical profile of human remains is huge, but we do know there is variation in the chemical profile. There is variation depending on whether a person is alive or dead, the stage of decomposition, variation depending on species, and, maybe, most importantly, variation in the environment in which the decedent is exposed. A doctoral dissertation by DeGreeff (2010) provided a graphic summary of some of these differences.

First, the chemical profile was found to differ between live and deceased humans. Although in black and white, Figure 4.11 shows the differences between living (F1, F2, F4, M1, M2, M3, and M4) and deceased (Mor1, Mor2, Mor3, Cr1, Cr2, and Cr3) humans. Although the quantities of chemicals collected may vary (seen by differences in the widths of the bands), there is a lot of similarities between the decedents' odor chemical profiles.

The odor profile differs depending on the stage of decomposition. The following summary by Degreeff (2010) describes the odor associated with different stages of decomposition:

Stage 1.) Fresh: Few exterior changes to the body, including the paling of the skin that results from the lack of oxygenated blood and the appearance of algor, livor, and rigor mortis. Internal decomposition begins as a result of bacterial action and autolysis. Stage one is also characterized by early skin slippage caused by autolysis at the dermal epidermal junction. A subtle odor is detectable by canines, but not by humans.

Stage 2.) Bloating or Putrefaction: The body creates an anaerobic environment that favors bacterial growth in the gut and bowel. Bacteria break down larger molecules causing the body to swell because of the internal production of gases. Black discoloration or a greenish tint may appear beneath the skin. The black discoloration of the skin is caused by the formation of a black precipitate during the breakdown of hemoglobin. A greenish discoloration is the result of the

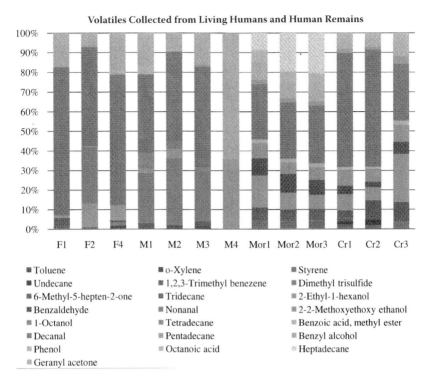

Figure 4.11 Difference in specific chemicals detected from living (F1, F2, F4, M1, M2, M3, M4) and deceased humans (Mor1, Mor2, Mor3, Cr1, Cr2, C3). (Courtesy of Lauryn DeGreeff, PhD disser., Florida International University. With permission.)

deamination of amino acids. During stage two, odor is detectable by both canines and humans.

Stage 3.) Decay: The skin ruptures releasing the gases that have built up inside, reintroducing oxygen to the body. This stage is the height of odor production and soft tissue loss.

Stage 4.) Liquefaction: The body begins to lose its integrity as the organs liquefy and the bones become visible. The body begins to dry out and odor is reduced.

Stage 5.) Skeletonization: The body's decay rate slows greatly as the last of the soft tissue decays. A slight, musty odor remains for some time. Bone may remain intact for many years or it may slowly be broken down by decalcification, dissolution by acid, or by scavenger activity.

Figure 4.12 shows the differences of chemical profiles between stage 1 and stage 2 decomposition. There are not a lot of differences between the types of chemicals, only some differences in the quantity.

Figure 4.13 shows the variation between different species. Again the graph makes it possible to appreciate the similarity (although not the quantity) of

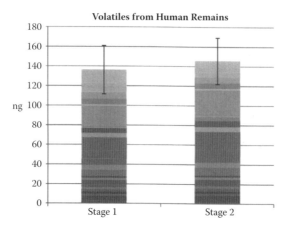

Figure 4.12 Chemical differences between stages of decomposition of the human. (Courtesy of Lauryn DeGreeff, Florida International University. With permission.)

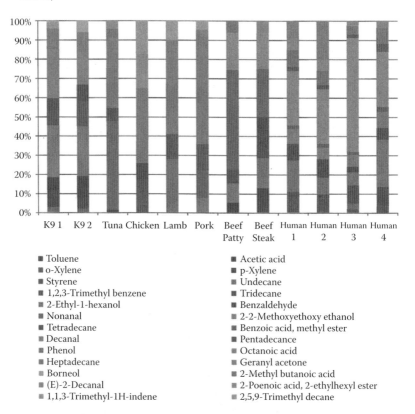

Figure 4.13 Variation in chemicals detected from various species and food products. K9 1 and K9 2 refer to deceased dogs while Human 1–4 refers to human decedents. (Courtesy of Lauryn DeGreeff, Florida International University. With permission.)

the four human remains samples contrasted with the others (K9 1 and K9 2 [deceased dogs] and various types of "meats").

Larson, Vass, and Wise (2011) described that key chemicals coming from a "freshly buried body") (<1285 ADD) could include sulfur dioxide, carbon tetrachloride, dimethyl-disulfide, toluene, benzene, dimethylbenzene (1,2 or 1,4), and freons (such as dichlorodifluoromethane). With burials with skeletonized decedents (>1285 ADD), chemicals could include "freons, aldehydes, such as nonanal, decanal," etc. Probably most significant is that the chemical profile of human remains likely varies depending on the location of the decedent and the environmental conditions encountered after death.

Thus, the message from this chapter is that, in many cases, the body breaks down with the help of a vast microbial population, creating and liberating a number of different chemicals. At this time, we do not know exactly what makes up the human remains (HR) chemical profile, but there have been a number of chemicals detected that are degradation products of carbohydrates, fats, and proteins. We do not know if these chemicals are the ones being detected by the HRD dogs, but can only assume so.

Once consensus is reached in the scientific community about the chemical profile of human remains, it will hopefully be possible to develop an "orthogonal" analytical device that complements the HRD dog (Figure 4.14).

Figure 4.14 Human remains detection (HRD) dog at work.

Any HR K9 handler can describe the frustration of being on the job and knowing the dog is in target odor, but is not able to pinpoint the location. As science and research move forward and the chemical profile is defined, the worlds of biological and mechanical sciences will hopefully get closer together.

Summary of Studies

The following abstracts and/or summaries were from studies that were done to help characterize the chemical profile of human remains. As previously demonstrated, there is a great degree of variation in the taphonomic process and the resulting chemical profile due to different conditions, temperature, water, microbial populations, etc. Because of this variation and the lack of consensus in the scientific community of the "exact" profile, it was decided to list these studies by publication date. As previously pointed out, this is *not* an all-inclusive list of all the research done. If any pivotal information was deleted, it was not the intention of the author.

> **Vass, A. A., W. M. Bass, J. D. Wolt, J. E. Foss, and J. T. Ammons.** 1992. Time since death determinations of human cadavers using soil solution. *Journal of Forensic Science* 37 (5): 1236–1253.
>
> This study was designed to collect data on specific volatile fatty acids, anions, and cations from soil underneath human bodies that were allowed to decompose naturally in Tennessee. The intent of the study was to develop information that could be used to help determine time since death or TSD.
>
> Soil was collected every three days in the spring and summer and weekly in the fall and winter from seven decedents. Based on information that the body is 40–50% muscle that is broken down by anerobic microbial fermentation to short-chain carboxylic acids (fatty acids), the soil was assayed. The soil in the wooded area had high organic matter content which likely bound with the purge fluid. The analysis was done on soil solution (liquid phase between soil particles). The authors reported the liberation of volatile fatty acids (VFAs) at an accumulated degree day temperature of 1285°C (± 110°C), so if a decedent was discovered and still had soft tissue present, the TSD could be calculated. The authors also reported that decedents exposed to similar environments also would "produce the same ratio of propionic, butyric, and valeric acids for any given ADD and that skeletonized remains also will show similar ratios of sodium, chloride, ammonia, potassium, calcium, magnesium, and sulfate for all given ADDs."

The authors reported that a large concentration of water-insoluble, long-chain fatty acids could indicate decomposition in a cold environment and that short-chain fatty acids are water soluble.

Vass, A. A., S. A. Barshick, G. Sega, J. Caton, J. T. Skeen, J. C. Love, and J. A. Synstelian. 2002. Decomposition chemistry for human remains: a new methodology for determining the postmortem interval. *Journal of Forensic Science* 47(3): 542–553.

This paper described the characterization of chemicals associated with human remains with the objective of identifying time-dependent biomarkers of decompositions. This work was done to develop an accurate and precise method for measuring the postmortem interval of human remains. Eighteen subjects were studied and tissues collected until the tissues decomposed to the point where they were no longer recognizable (a range of approximately three weeks). Analysis of collected tissues (liver, kidney, heart, brain, and muscle) for specific biomarkers revealed "distinct patterns useful for determining the PMI (postmortem interval)."

Lorenzo, N., T. L. Wan, R. J. Harper, Y. L. Hsu, M. Chow, S. Rose, S., and K. G. Furton. 2003. Laboratory and field experiments used to identify *Canis lupis* var. familaris active odor signature chemicals from drugs, explosives, and humans. *Analytical and Bioanalytical Chemistry* 376: 1212–1224.

This paper described the use of laboratory analysis to identify the signature odors that law enforcement-certified detector dogs alert to when searching for drugs, explosives, and humans. Studies included the analysis and identification of the "fingerprint" of a variety of samples, followed by completion of double-blind dog trials of the individual components. This was an attempt to isolate and understand the target compounds that dogs alert to. Air surrounding the "human materials" were sampled, analyzed, and revealed "trimethylamine, 1-pentanol, hexanal, butanoic acid, pentanoic acid, heptanal, benzaldehyde, 2-pentryl furan, dimethyl disulfide, hexanoic acid, heptanoic acid, nonanoic acid, and octanoic acid. Major components seen in the headspace of pig decomposition materials were oleic acid, 2-anthracenamine, propanoic acid, butanoic acid, and hexadecanoic acid."

Vass, A. A., R. R. Smith, C. V. Thompson, M. N. Burnett, D. A. Wolf, J. A. Synstelian, N. Dulgerian, and B. A. Eckenrode. 2004. Decompositional odor analysis database. *Journal of Forensic Science* 49(4): 760–769.

Vass, A. A., R. R. Smith, C. V. Thompson, M. N. Burnett, N. Dulgerian, and B. A. Eckenrode. 2008. Odor analysis of decomposing buried human remains. *Journal of Forensic Science* 53 (2): 384–391.

The following is information from two articles authored by Vass et al. The studies were designed to identify the "chemical fingerprint" of buried human decedents (2004, 2008). Air samples were collected below and at the surface of decedents buried from 1.5 to 3.5 feet deep. Samples were collected about once a week in the summer and monthly in the winter. The goal was to identify the chemical signature to help develop training aids for HRD dogs as well as the development of a portable analytical instrument that could be used in the field.

Over 475 volatile or semivolatile chemicals were identified, all known products of soft tissue decomposition. Very light volatiles, including ammonia, methane, hydrogen, and carbon dioxide, were not included. There were eight different classes that were separated on the basis of abundance, and then correlated with environmental factors. These were then ranked on the most significant and persistent VOCs. Vass identified 30 compounds as key markers detected at the soil surface that moved up unchanged from the body.

By organizing or ranking the VOCs by burial accumulated degree days (BADD), Vass et al. developed a timeline of emissions, possibly providing some insight into how the HR odor profile changes over time. Table 4.8 shows three different groups of compounds detected at three different times.

Inoue, H., M. Iwasa, Y. Maeno, H. Koyama, Y. Sato, and R. Matoba. 1996. Detection of toluene in an adipoceratous body. *Forensic Science International* 78: 119–124.

A 24-year old decedent was found in a car in a river. It was believed that the decedent had been drinking prior to death. As no remarkable injuries were seen and diatoms were found in the lungs, it was assumed that the decedent drowned about three months prior to being found. Toluene was detected in liquid from a bottle in the car. The body was described as "almost adipoceratous."

Detectable levels of toluene were found in the decedent's brain, liver, kidneys, smooth muscle, and adipose tissue. The following chemicals were also detected: ethanol, 1-propanol, 2-propanol, 1-butanol, dimethyldisulfide, isovaleraldehyde, and n-butyl-n-butyrate.

Statheropoulos, M., C. Spiliopoulou, and A. Agapiou. 2005. A study of volatile organic compounds evolved from the decaying human body. *Forensic Science International* 153: 147–155.

Two decedents were found in Greece in seawater with an estimated postmortem interval (PMI) of three to four weeks. They were moderately to advanced putrefied. Air samples were collected from inside the body bags and included samples from a body bag without a body. Sampling took place at a temperature of 14°C and

Table 4.8 Different groups of chemicals varying by decomposition age*

Group 1	Group 2	Group 3
Present throughout the decomposition of both soft tissue and bone (out to 16 years postburial). Least cyclic.	Includes early decomposition for up to one year postburial. Most cyclic of three groups. BADD < 7300.	Present as long as soft tissue present. BADD < 18,000.
Benzene derivatives, halogenated compounds, aldehydes.	Esters, benzene derivatives, and some halogens.	Many sulfur and halogenated compounds.
Ethyl benzene	Trichloroethene	Dichlorodifluoromethane
Toluene	1-methyoxypropyl benzene	Dimethyldisulfide
Tetrachloroethene	Sulfur dioxide	Ethane,1,1,2-trichloro-1,2,2-trifluoro
1,4-dimethylbenzene	Hexadecanoic acid, methyl ester	Dimethyltrisulfide
Carbon tetrachloride	Hexane	Chloroform
1,2-dimethylbenzene	1,1-dichloro-1-monofluoroethane	
Naphthalene	1-ethyl,2-methylbenzene	
Styrene	Benzenemethanol, α, α-dimethyl	
Benzene	Methenamine	
Nonanal	1,2-benzenedicarboxylic acid, diethyl ester	
Decanal		
Trichloromonofluoromethane		
Calcium disulfide		
Undecane		

* Adapted from Vass, 2004.

52% humidity (case 1) and 17°C and 52% humidity (case 2). In this study, 60 chemicals were collected, with 15 in common at similar concentrations from the two bodies. The list of common chemicals included: dimethyldisulfide, toluene, hexane, 1,2,4-trimethylbenzene, 2-propanone, 3-pentanone, 2-pentanone, and 2-methylpentane. Toluene levels were high in both decedents, possibly due to premortem exposure. Cadaverine and putrescine were not detected.

Eckenrode, B.A., S. A. Ramsey, R. A. Stockham, G. J. VanBerkel, K. G. Asano, and D. A. Wolf. 2006. Performance evaluation of the scent transfer unit (STU-100) for organic compounds collection and release. *Journal of Forensic Science* 51 (4): 780–789.

This study involved evaluation of a portable vacuum system used to collect air samples from various sources. This device was evaluated to determine its ability to trap and release organic compounds at ambient temperature under controlled laboratory conditions. Testing of the STU-100 was found to have significant trapping efficiency at ambient temperatures.

Lesniak, A., M. Walczak, T. Jezierski, M. Sachurczuk, M. Gawkowski, and K. Jazscsak. 2008. Canine olfactory receptor gene polymorphism and its relation to odor detection, performance by sniffer dogs. *Journal of Heredity* 99 (5): 518–527.

Lesniak et al. described work investigating specific genes controlling dog olfactory receptors. After testing the performance of five different types of detector dogs, the authors concluded their findings suggested "a role of specific alleles in odor detection and a linkage between single-nucleotide polymorphism and odor recognition efficiency." In summary, there may be a genetic predisposition in some dogs for effective detector work.

Oesterhelwig, L., S. Krober, K. Rottman, J. Willhoft, C. Braun, N. Thies, K. Puschel, J. Silkenath, and A. Gehl. 2008. Cadaver dogs: A study on detection of contaminated carpet squares. *Forensic Science International* 174: 35–39.

Carpet squares were contaminated with "scent of two recently deceased bodies" whose postmortem interval was less than three hours. There was a light cotton wrap between the body and carpet squares that were in contact with the bodies for either 2 minutes or 10 minutes. Dogs were able to detect odor up to 35 days on the 2-minute exposure carpet and up to 65 days on the 10-minute exposure carpet. The authors could not find a decrease in "accuracy of dogs' performance over age of target odor."

Statheropoulos, M., A. Agapiou, C. Spiliopoulou, G. C. Pallis, and E. Sianos. 2007. Environmental aspects of VOCs evolved in the early stages of human decomposition. *Science of the Total Environment* 385: 221–227.

The study involved a human decedent who had been dead for four days and then was placed in a sealed bag. Decomposition was not "progressed" at that time. Air samples were collected at 0, 4, 8, and 24 hours. Thirty VOCs were collected, with most of them found at the 24-hour sampling timepoint. The major families included many benzene derivatives, aliphatic hydrocarbons, oxygenated compounds (alcohols, aldehydes, ketones, and sulfides). No diamines (cadaverine or putrescine) were found. Many gases also were found, including H_2S, CO_2, CH_4, NH_3, SO_2, and H_2.

The authors pointed out that there were compounds found in both the 2004 Vass paper and theirs that included: heptanes, 2-propanone, dimethyldisulfide, methylbenzene, xylene (p, o, and m).

Dekeirsschieter, J., F. J. Verheggen, M. Gohy, F. Hubrecht, L. Bourguignon, G. Lognay, and E. Haubruge. 2009. Cadaveric volatile organic compounds released by decaying pig carcasses (*Sus domesticus* L.) in different biotopes. *Forensic Science International* 189: 46–53.

The authors studied the decompositional VOCs released by decaying pig carcasses. Cadaveric VOCs released by pig carcasses decomposing in three biotypes (crop field, forest, urban site) were collected and analyzed. The authors reported that there were no cadaveric VOCs detected during the "fresh decompositional stage." There were 90, 85, and 57 cadaveric VOCs identified from pig carcasses laying on the agricultural site, forest biotope, and the urban site, respectively. The main cadaveric VOCs were acids, cyclic hydrocarbons (indole, phenol, and 4-methylphenol), oxygenated compounds, sulfur, and nitrogen compounds. During advanced decay when soft tissues were removed, the "portion" of aldehydes increased.

Of the total 104 VOCs identified, 35 were found in all three environmental settings. These included 7 acids, 4 esters, 1 ketone, 1 aldehyde, 5 alcohols, 6 N compounds, 5 sulfur compounds, 4 cyclic hydrocarbons, and 2 noncyclic hydrocarbons.

Hoffman, E. M., A. M. Curran, N. Dulgerian, R. A. Stockham, and B. A. Eckenrode. 2009. Characterization of the volatile organic compounds present in the headspace of decomposing human remains. *Forensic Science International* 186: 6–13.

Hoffman et al. published results from a study done to characterize the VOCs from a number of tissue types that are "commonly used" as training aids for HRD dogs. The tissues analyzed in this study included human blood (clot), human placental blood, human blood, muscle, testicle, skin, body fat attached to skin, adipocere (two samples), fat, bone (vertebrae), bone (two samples), and teeth. The tissues were placed in glass vials at about 45°F (7°C), at 45% humidity, and allowed to "warm up" for 4 to 48 hours before the air above the tissues were analyzed. Among the 14 tissue samples, 33 common VOCs were identified within 7 chemical families. No two tissues generated the same profile, but there were qualitative similarities and differences. When similar type tissues were grouped together, a number of VOCs were found that included (tissue/number VOCs): fat/22; muscle/19; bone/19; adipocere/18; blood/12. Hexanol was found in blood, muscle, adipocere body fat with skin and bone (all three samples). Dimethyldisulfide also was found in most of the tissues sampled, including testicle, adipocere, skin, placental clot, fat and skin, adipocere, vertebrae, bone (two), and teeth (three total). It is important to

Table 4.9 Number of chemicals in each class detected from "isolated" human tissues[*]

Tissue	Acid Esters	Alcohol	Aldehydes	Aromatic Hydrocarbons	Halogenated Compounds	Ketones	Sulfides
Fat	5	5	10	2	—	—	—
Fat and Skin	4	4	2	3	—	1	1
Adipocere 1	1	1	—	3	1	—	—
Adipocere 2	4	2	7	3	—	1	1
Vertebrae	—	2	1	2	1	1	1
Bone 1	—	1	4	2	1	—	1
Bone 2	2	5	9	2	—	1	—
Muscle	3	4	9	2	—	1	—

[*] Adapted from Hoffman et al. (2009).

note that these samples did not include surface or subsurface factors and no soil microbes (Table 4.9).

Hudson, D. T., A. M. Curran, and K. G. Furton. 2009. The stability of collected human scent under various environmental conditions. *Journal of Forensic Science* 54 (6): 1270–1276.

Human scent evidence collected from objects at a crime scene can be used for scent discrimination with specially trained canines. Storage of the scent evidence is usually required yet no optimized storage protocol has been determined. Storage containers for scent evidence including glass, polyethylene, and aluminized pouches were evaluated to determine the optimal medium for storing human scent evidence. Glass was shown to be the best container to store human scent samples in. The study also showed that the glass jars should be kept out of sunlight (ultraviolet A (UVA) and ultraviolet B (UVB)).

Curran, A. M., P. A. Prada, and K. G. Furton. 2010. The differentiation of the volatile organic signatures of individuals through SPME-GC/MS of characteristic human scent compounds. *Journal of Forensic Science* 55 (1): 50–57.

This study described the development of what is effectively a human scent barcode or "fingerprint" made up of different amounts of primary odor. This individual and unique profile could be stored in a searchable database, providing another way for identification of individuals. By narrowing the compounds considered for each subject to only those common in all three samples, the individuals were correctly discriminated and identified in 99.54% of the cases.

References

Anon. 2006. Organic chemistry study guide: Organic compounds, formulas, isomers, nomenclature (v. 12.1). Boston: MobileReferences Publishing.

Brisson, A, and D. Nibbe. 2003. *Beer guide: Quick study*. Boca Raton, FL: BarCharts, Inc.

Bloch, D. R. 2006. *Organic chemistry demystified: A self teaching guide*. New York: McGraw-Hill Publishing.

Cablk, M. E., E. E. Szelagowski, and J. C. Sagebiel. (2012). Characterization of the volatile organic compounds present in the headspace of decomposing animal remains, and compared with human remains. *Forensic Science International* in press.

Carpi, A. n.d. *Organic chemistry*. Online at: www.visionlearning.com (accessed July 7, 2011).

Curran, A. M., P. A. Prada, and K. G. Furton. 2010. The differentiation of the volatile organic signatures of individuals through SPME-GC/MS of characteristic human scent compounds. *Journal of Forensic Science* 55 (1): 50–57.

DeGreeff, L. 2010. Development of a dynamic headspace concentration technique for the non-contact sampling of human odor samples and the creation of canine training aids. PhD diss., Florida International University.

Dekeirsschieter, J., F. J. Verheggen, M. Gohy, F. Hubrecht, L. Bourguignon, G. Lognay, and E. Haubruge. 2009. Cadaveric volatile organic compounds released by decaying pig carcasses (*Sus domesticus* L.) in different biotopes. *Forensic Science International* 189: 46–53.

Eckenrode, B. A., S. A. Ramsey, R. A. Stockham, G. J. VanBerkel, K. G. Asano, and D. A. Wolf. 2006. Performance evaluation of the scent transfer unit (STU-100) for organic compounds collection and release. *Journal of Forensic Science* 51 (4): 780–789.

Forbes, S. L. 2008. Decomposition chemistry in a burial environment. In *Soil Analysis in Forensic Taphonomy: chemistry and biological effects of buried human remains*. Eds. M. Tibbett and D. O. Carter. Boca Raton FL: CRC Press/Taylor and Francis, Chap. 8.

France, D. L. 2010. *Human and nonhuman bone identification: A concise field guide*. Boca Raton, FL: CRC Press/Taylor & Francis Group.

Hart, H. 1983. *Organic chemistry: A short course,* 6th ed. Dallas: Houghton Mifflin Company.

Harvey, L. M., S. J. Harvey, M. Hom, A. Perna, and J. Salib. 2006. The use of bloodhounds in determining the impact of genetics and the environment on the expression of human odor type. *Journal of Forensic Science* 51 (5): 1109–1114.

Hawari, J., S. Beaudet, A. Halasz, S. Thiboutot, and G. Ampleman. 2000. Microbial degradation of explosives: Biotransformation versus mineralization. *Applied Microbiology and Biotechnology* 54: 605–618.

Hoffman, E. M., A. M. Curran, M. Dulgerian, R. A. Stockham, and B. A. Eckenrode. 2009. Characterization of the volatile organic compounds present in the headspace of decomposing human remains. *Forensic Science International* 186: 6–13.

Hole, J. W. 1990. *Human anatomy and physiology*. Dubuque, IA: Wm. C. Brown Publishing.

Hudson J. T., A. M. Curran, and K. G. Furton. 2009. The stability of collected human scent under various environmental conditions. *Journal of Forensic Science* 54 (6): 1270–1277.

Inoue, H., M. Iwasa, Y. Maeno, H. Koyama, Y. Sato, and R. Matoba. 1996. Detection of toluene in an adipoceratous body. *Forensic Science International* 78: 119–124.

Jackson, M. 2001. *Organic chemistry fundamentals: Quick study guide.* Boca Raton, FL: BarCharts, Inc.

Larsen, D. O., A. A. Vass, and M. Wise. 2011. Advanced scientific methods and procedures in the forensic investigation of clandestine graves. Journal of Contemporary Criminal Justice XX (X): 1–34.

Lesniak, A., M. Walczak, T. Jezierski, M. Sachurczuk, M. Gawkowski, and K. Jazscsak. 2008. Canine olfactory receptor gene polymorphism and its relation to odor detection, performance by sniffer dogs. *Journal of Heredity* 99 (5): 518–527.

Lorenzo, N., T. L. Wan, R. J. Harper, Y. L. Hsu, M. Chow, S. Rose, and K. G. Furton. 2003. Laboratory and field experiments used to identify *Canis lupis* var. familaris active odor signature chemicals from drugs, explosives, and humans. *Analytical and Bioanalytical Chemistry* 376: 1212–1224.

Metting Jr., F. B., ed. 1993. Structure and physiological ecology of soil microbial populations. In *Soil microbial ecology: Applications in agricultural and environmental management.* New York: Marcel Dekker, Chap. 1.

Miller, F. C. 1993. Composting as a process based on control of ecologically selective factors. In *Soil microbial ecology: Applications in agricultural and environmental management,* ed. F. G. Metting. NY: Marcel Dekker, Chap. 8.

Owen, J. 2005. Cow power: Battery runs on bovine stomach bacteria. *National Geographic News,* September 9.

Oesterhelwig, L., S, Krober, K. Rottman, J. Willhoft, C. Braun, N. Thies, N., K. Puschel, J. Silkenath, and A. Geth. 2008. Cadaver dogs: A study on detection of contaminated carpet squares. *Forensic Science International* 174: 35–39.

PhysOrg. n.d. *Forensic chemists verify human remains from fat deposits.* Online at: www.physorg.com/news144955495.html (accessed September 12, 2011).

Ruttan, R. F., and M. J. Marshall. 1917. *The composition of adipocere.* Online at: www. jbc.org (accessed June 20, 2011).

Statheropoulos, M., C. Spiliopoulou, and A. Agapiou. 2005. A study of volatile organic compounds evolved from the decaying human body. *Forensic Science International* 153: 147–155.

Statheropoulos, M., A. Agapiou, C. Spiliopoulou, G. C. Pallus, and E. Sianos. 2007. Environmental aspects of VOCs evolved in the early stages of human decomposition. *Science of the Total Environment* 385: 221–227.

Takatori, T. 1996. Investigations on the mechanism of adipocere formation and its relation to other biochemical reactions. *Forensic Science International* 80: 49–61.

Thompson, A. 2007. *Scientists ruminate on cow stomach fluid for fuel cells.* Online at: www.livescience.com (accessed July 7, 2011).

Vass, A. A., W. M. Bass, J. D. Wolt, J. E. Foss, and J. T. Ammons. 1992. Time since death determinations of human cadavers using soil solution. *Journal of Forensic Science* 37 (5): 1236–1253.

Vass, A. A., S. A. Barshick, G. Sega, J. Caton, J. T. Skeen, J. C. Love, and J. Synstelian. 2002. Decomposition chemistry for human remains: A new methodology for determining the postmortem interval. *Journal of Forensic Science* 47(3): 542–553.

Vass, A. A., R. R. Smith, C. V. Thompson, M. N. Burnett, D. A. Wolf, J. Synstelian, N. Dulgerian, and B. A. Eckenrode. 2004. Decompositional odor analysis database. *Journal of Forensic Science* 49: 760–769.

Vass, A. A., R. R. Smith, C. M. Thompson, M. N. Burnett, N. Dulgerian, B. A. Eckenrode, 2008. Odor analysis of decomposing buried human remains. *Journal of Forensic Science* 53 (2): 384–391.

Wilson, A. S., R. C. Janaway, A. D. Holland, H. I. Dodson, E. Baran, and A. M. Pollard. 2007. Modeling the buried human body environment in upland climes using three contrasting field sites. *Forensic Science International* 169: 6–18.

Earth, Wind, and Odor

5

Introduction

This is the chapter that hopefully pulls information together from the previous chapters to help you understand how truly amazing that our human remains detection (HRD) dogs can do what they do. Although this chapter will not provide you with precise answers about the environmental dispersion of all those yet undefined chemicals that make up the signature of human remains. It should help to explain or reinforce some of the concepts of how we can use our limited knowledge of the environment to explain (in part) how HRD dogs work (Figure 5.1).

So, at this point, it is necessary to try and take the information summarized in Chapter 4 and apply it. This may take a stretch of the imagination because there are numerous assumptions made, and this is an area of science that is still rather new. Obviously we don't know if this information is representative of the chemical profile of human remains or if this is what the dog actually keys in on, but a handler should explain to others what and how the HRD K9 works. That is a large order, so let's begin.

Chemistry: Highlights of Chapter 4

Before we can truly understand the environmental effects on odor dispersion, it is necessary to explain some of the physicochemical properties of some of the chemicals associated with human remains.

Gas (or vapor) basics:

- Gas can convert to a liquid form if there is enough pressure to compress the molecules together squeezing it into a liquid.
- Generally, as temperature increases, gas molecules become more energetic, making it more difficult for it to liquefy.
- Critical temperature is defined as the temperature (at or above) where gas cannot be liquefied, no matter how much pressure it is put under.

Figure 5.1 HRD dogs are ready to work, be it on land or in the water.

- Critical pressure is the amount of pressure required to liquefy a gas at its critical temperature (Purdue University, n.d.).

Table 5.1 shows the differences in *critical temperature* and *critical pressure* of some common chemicals/compounds. It should help to understand the effects of critical temperatures and pressure on these compounds. First, temperature. For H_2O, it takes a higher temperature (and energy) to turn liquid to gas. The critical temperatures for O_2 (oxygen gas) or CO_2 (carbon dioxide) are much lower.

Pearsall and Verbruggen (1982) described the ability of a chemical to give off an odor based on its boiling point. As the boiling point decreases, the evaporation and dispersion of a chemical increases.

Table 5.1   Critical temperatures and pressures of common chemicals and/or compounds

Chemical	Critical Temperature (°C)	Critical Pressure (atm)
Ammonia (NH_3)	132.0	111.5
O_2	−119.0	49.7
CO_2	31.2	73.0
H_2O	374.0	217.7

Next—pressure. The *vapor pressure* (VP) of a chemical or compound is the pressure at which a vapor or gas is "thermodynamically" equal to its condensed form or phase in a closed system. Which means: Put a chemical in a closed container that is pressurized and vapor pressure is what it takes to make the gas phase equal the liquid phase.

The VP of a chemical can be an indication of a liquid's evaporation rate. A chemical that has a high VP at normal temperature (usually considered to be around 20 to 25°C) is called a volatile chemical. VP is measured as a standard unit of measure at a given temperature. For this book, VP is measured in millimeters of mercury [mmHg] at 25°C. The VP of water is 17.5 mmHg @ 20°C; VP of O (oxygen) is 407,936 mmHg @ 20°C, and VP of N (nitrogen) of 475,106 mmHg @ 20°C. Remember, the higher the VP, the more volatile it is.

Henry's Law Constant is another measurement that describes the volatility of a chemical in the environment. It is defined as the solubility of a gas suspended (or trapped) in a liquid at a given temperature that is proportional to the pressure of the gas phase above the liquid phase (Purdue University, n.d.). A commonly used example of Henry's Law Constant has to do with carbonated soft drinks, called "soda" or "pop" depending on what part of the country you are from (Figure 5.2). Before the can or bottle is opened, the gas above the liquid soda is almost pure carbon dioxide (CO_2), which is at a pressure slightly above atmospheric pressure. Once the can or bottle is opened, some CO_2 gas escapes ("pops"), decreasing pressure inside the can. The decrease in pressure causes the liquid CO_2 in the soda to volatilize into those little gas bubbles that we all enjoyed as a kid. If there is an equal amount of CO_2 gas and CO_2 liquid in the soda, it becomes flat. So, Henry's Law describes the actual pressure at which soda would become flat or when the gas phase and liquid phase of a given chemical are equal. There

Figure 5.2 Henry's Law Constant = solubility of a gas suspended (or trapped) in a liquid at a given temperature that is proportional to the pressure of the gas phase above the liquid phase.

Figure 5.3 Oil and water do not mix. (Courtesy of J. K. Rosenbaum.)

are different units for this, but we will use atmospheric pressure per cubic meter of air at 25°C or atm/m3@25°C.

Another term to understand is a *partition coefficient*. This is basically what Henry's Law is all about, except it also can be used to describe the ratio of concentrations of a compound when it is dissolved in a mixture of two different "immiscible solvents," which are ones that don't mix, such as oil and water (Figure 5.3). Chemists often use water as a solvent that attracts hydrophilic or water-loving chemicals. The "oil"-like solvent is often octanol, a hydrophobic (water-fearing) solvent. Octanol/water partition coefficient (Kow) is a relative measurement of how water-loving or water-fearing a chemical is. Kow gives an indication of how easily a chemical may get stuck or move into groundwater. Another type of partition coefficient is Koc—the ratio of how much of a chemical sticks to the organic carbon or moves into water. Kow describes how a chemical may stick to soil or sediment.

Table 5.2 includes some of the chemicals associated with human remains. Although many others have been identified, this list includes some of the chemicals that have known physicochemical values that can be used to predict how these may move through the environment. The table contains information taken from the Hazardous Substances Database maintained by the National Institutes of Health's National Library of Medicine (NLM).

So, what does this really mean for a given chemical?

High vapor pressure = highly volatile
Low Henry's Law Constant = highly volatile
Low Koc = highly mobile in soil
High log Koc = high sorption to soil or sediment
Low Kow = moves into water
High Kow = sticks in environment

Table 5.2 Physicochemical characteristics of some chemicals associated with human remains

Chemical	Benzaldehyde	Ethyl n-butyrate	Heptanoic acid	Methane	Methyl disulfide	Nonanal	Oleic acid	Toluene	Trichloroethylene
Class	Aldehydes	Acid ester	Acid	Alkane	Sulfur		Fatty acid	Cyclic hydrocarbon	Alkene
Vapor Pressure (mmHg@25°C)	0.127	12.8	$1.07 \times 10{-2}$	$4.66 \times 10{+5}$	28.7	$3.7 \times 10{-1}$	$5.46 \times 10{-7}$	28.4	69
Henry's Law (atm/m3/mole)	$2.67 \times 10{-5}$	$3.99 \times 10{-4}$		$6.58 \times 10{-1}$	$1.21 \times 10{-3}$	$7.34 \times 10{-4}$	$4.48 \times 10{-5}$	$6.64 \times 10{-3}$	$9.85 \times 10{-3}$
Phase in ambient air	Vapor	Vapor	Vapor	Vapor	Vapor	Vapor	Vapor and particulate	Vapor	Vapor
Photochemical degradation (half-life)	2–16 days	6 days	2.3 days	6 years	4 hours (photolytic)	12 hours	Vapor phase degraded in ~6 hours	3 days	7 days

Atmospheric Fate

(continued)

Table 5.2 Physicochemical characteristics of some chemicals associated with human remains (continued)

Chemical	Benzaldehyde	Ethyl n-butyrate	Heptanoic acid	Methane	Methyl disulfide	Nonanal	Oleic acid	Toluene	Trichloroethylene
Koc	34	41	490	90		1600	340,000	37–178	101
logKow		1.73	2.42	1.09	1.77		7.64		
Mobility in soil	High	High	Moderate	High		Low	Immobile	Moderate to high	High
Terrestrial Fate									
Soil sorption	NE	NE	Moderate		High		High		Low
Volatilization				High					
Wet soil	High		Low		High	High	None	High	Very high
Dry soil	Moderate		NE		High	High	None	High	Moderate
Degradation	High		Moderate			High	High	High	
Aerobic									None
Anaerobic									Low
Aquatic Fate									
Sorb to suspended/sediment	Not expected	No	Moderate	No		Moderate	High	No	No
Volatilize from:	High	High	NE	High	High	High		Moderate	High
River	1.5 days	5.5 hours			3 hours	5 hours		1 hour	3.5 hours
Lake	14 days	65 hours			4 days	5 days		4 days	5 days
Bioconcentration	Low	Low			Low	High	Low		Low

Source: Hazardous substances database. Bethesda, MD: National Institutes of Health's National Library of Medicine.

Although these chemicals have different physicochemical values, it is important to remember that:

- The human remains (HR) odor is a mixture of chemicals, some of which may or may not be present at different times, under different conditions, and at different concentrations.
- Assuming the remains stays in one place, it should continue to off gas and replenish any odor chemicals that may be removed or changed by environmental factors.
- Environmental factors can affect how and what chemicals may be present and disperse to be detected by the HRD dog.

Therefore, what this part shows that, in certain situations, some chemicals can volatilize to the atmosphere, some of the chemicals may adsorb or stick to things in the environment (soil, plants, etc.), and that some of the chemicals may be picked up and transported in water. This information provides a background for the handler to understand that HR odor dispersion is very complex, probably as complex as the taphonomic process that it took to create this chemical profile. It is now time to use this information to explain what the dog's nose knows.

The Dog's Nose Knows

Much of the following is taken from *The Cadaver Dog Handbook: Forensic Training and Tactics for the Recovery of Human Remains* (Rebmann, David, and Sorg, 2000). This section will explain the effects of weather, topography, etc., on the dispersion of HR odor, something these authors captured in their diagrams and what dogs have been showing us for years. We will get to those figures soon, but first it is necessary to summarize their principles of scent cone presence and distortion. As an HRD dog handler, I can attest to the accuracy of these drawings and principles. My dogs have shown me this many times throughout the past decade of work. The modified, summarized principles include:

- Decomposition odor forms a scent pool around the decedent (primary scent pool). The scent molecules gradually "shed into the air and are absorbed into the soil in all directions."
- Thermal uplift of the scent pool occurs as air currents rise due to warming, causing a vertical movement of the odor.
- Air flow (wind) will move scent away from the primary scent cone in the direction of the wind, forming a horizontal scent cone. The

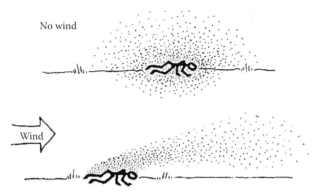

Figure 5.4 Odor cone around body when there is no wind. (From Rebmann, A., E. David, and M. H. Sorg. 2000. *The cadaver dog handbook: Forensic training and tactics for the recovery of human remains.* Boca Raton, FL: CRC Press. With permission.)

faster the wind, the narrower the cone; kind of like a water hose with a constricted nozzle (Figure 5.4).

- Variable wind patterns can cause an uneven distribution of scent molecules in the air, possibly causing breaks or void in the scent.
- If a body is elevated and has a horizontal scent cone, there can be scent voids near the remains at "dog-nose levels."
- Water will move scent away from the remains along "scent conduits" in response to gravity and/or currents, along surface or underground waterways, or following runoff, erosion, or drainage patterns.
- Water flow along a conduit can interrupt the absorption of scent into soil near the remains, causing a scent void near the remains at dog-nose level.
- Wind or water flow can be altered by barriers (e.g., vegetation, buildings, rock, hills, etc.), which can cause the formation of secondary scent pools and, potentially, forming new secondary scent cones away from the remains.

With a list like that, it may be hard to visualize what each point may actually look like. We will get back to those after we cover some of the basics of the environmental variables.

Weather or Not

Figure 5.4 shows the odor cone around a body in a neutral environment and being dispersed by wind. The volatile chemicals off gas and remain in the air over and around the body. What we do know is that this kind of environmen-

Figure 5.5 Atmosphere is a mixture of gases, water, dust, and other particulates held in place with gravity.

tal scenario is pretty unusual and that things like temperature, wind, etc., alter and influence odor dispersion. Let's start with basic weather concepts.

Atmosphere

Atmosphere is basically defined as a mixture of gases, water, dust, and other particulates held in place by gravity (Figure 5.5). As you may remember from Earth Science class, the atmosphere has several different layers, starting with the troposphere, the layer closest to the earth and really the one with which we are most concerned. The troposphere insulates, protects, and sustains life and contains most of our weather. (Much of the information in this section is from Herd (2001).)

The troposphere contains 78.1% nitrogen, 20.9% oxygen, 0.93% argon, and about 0.04% carbon dioxide, and anywhere from 0–4% water. The particulates in the troposphere air include dust and other matter, which can have a dramatic effect on local and short-term weather (Figure 5.6). An erupting volcano in Iceland in 2010 shut down air travel in most of Europe for several days (Weise, Vergano, and Rice, 2010); a pretty substantial effect on local and regional weather.

Atmospheric pressure is how much air weighs if it were contained in a column extending from the Earth's surface to space. It is usually measured by a barometer (Figure 5.7). Generally the heaviest and densest air is at the

Figure 5.6 Dust cloud from soil particulates picked up by the wind.

Figure 5.7 Barometer measures atmospheric pressure. (Courtesy of J. K. Rosenbaum.)

Earth's surface (carrying the weight of the total mass), and as you go higher, the air becomes lighter and thinner. Density and temperature are two parts of atmospheric pressure. Although the atmospheric layers are held in place by gravity, gases like to resist the pull of gravity and instead rise. Also, as you get higher in the atmosphere, the temperature generally decreases. So, when gas at the Earth's surface heats up and expands, it takes up more space with the same amount of molecules, and, therefore, decreases density. This causes

Thermal
uplift

Figure 5.8 Thermal lift of air containing human remains (HR) odor. (From Rebmann, A., E. David, and M. H. Sorg. 2000. *The cadaver dog handbook: Forensic training and tactics for the recovery of human remains.* Boca Raton, FL: CRC Press. With permission.)

the air to rise. Figure 5.8 shows this concept: The air containing volatile gases from the body rise along with the water (vapor)-containing air.

The temperature part of weather comes from the transfer of heat (energy) from one object to another. The transfer of electromagnetic *radiation* from the sun to a dark surface is one method of heat transfer. A common example of this type of energy transfer is a dark car parked in direct sunlight (Figure 5.9). Another type of heat transfer is *conduction*; this is generally seen more often in solids and liquids, and not so much with gases. The other type of energy (heat) transfer is convection, which is the transfer of heat by movement through a fluid or gas. One example of convection is a hot cup of coffee heating up the mug in which it is poured. Convection is also the upward movement of air up. Advection is the horizontal movement of energized air.

There are a couple of other terms that should be covered concerning temperature. *Solar radiation* is when electromagnetic (EM) waves from the sun bring energy in the form of heat to the Earth. The size and properties of gas particles in the atmosphere will affect how much scattering and absorption of

Figure 5.9 Electromagnetic radiation: Transfer of heat to dark car in the sun. (Courtesy of J. K. Rosenbaum.)

Figure 5.10 Terrestrial radiation: Heat reflected from lake.

energy takes place. If there are a lot of particles, etc., in the atmosphere (such as clouds), then less of the EM waves (heat) will make it to the Earth's surface.

Terrestrial radiation is what occurs when heat is reflected back from the Earth. An example is when water vapor in the air absorbs, captures, and holds heat from the Earth during the day (Figure 5.10). This water vapor will hold heat; that is why a hot, humid night doesn't really cool off.

Albedo is a term used to describe the radiation reflectivity of a surface. With the black car, with a reflectivity value of 0, most (if not all) of the solar EM waves are absorbed. However, with a white car with an albedo value of 1 (or 100%) equals less heat absorbed.

Proximity or angle of the sun also affects the solar radiation and heating of air. The greatest expansion and rising of air occurs near the equator because the sun does not shift up and down in the sky as much as in other geographical regions. This plays a role in the Coriolis effect or force where the dividing line of the equator causes air and water currents to circle to the right in the Northern Hemisphere and to the left in the Southern Hemisphere.

In summary: On a cool day, the odor will be close to the body and, as the day heats up, the air with the volatile chemicals will radiate up and out. Then as the air cools, the odor can settle away from the victim, creating a large and dispersed odor pool (Hammond, 2006).

Wind

Where does wind come from? As hot air rises and moves up and mixes in layers of air with cooler temperatures, it then sinks, causing vertical or

Figure 5.11 Hot air balloon. (Courtesy of PDPhoto.org.)

up-and-down movement of air. Air also moves horizontally, caused mainly by differences in pressure. This would be the high and low pressure that you see described on the Weather Channel or a local TV station. As learned in high school, nature abhors a vacuum, causing air from high pressure systems to flow to low pressure areas. The sideways air movement is wind. The rate of this air (or gas) cycling is called *adiabatic lapse rate*.

As you will remember, with free convection, locally heated air (which is warmer than the surrounding air) will rise like a hot air balloon. This is true especially for water vapor that weighs even less that either O or N (Figure 5.11). As air rises and expands (due to less gravity), the pressure decreases and then the molecule cools. As it cools, it sinks, and wind is created.

Topography

Odor dispersion also can be affected by location, terrain, and local barriers. Wind (measured by something called an anemometer) moves faster over open (flat) surfaces, and slower in areas with "texture," such as hills,

Figure 5.12 Any kind of barrier can affect wind speed and direction, much like rocks in a river can affect water flow.

buildings, vegetation, etc. (Figure 5.12). Differences in topography can cause distortion in air flow, something that any HRD dog handler can attest to.

During the day, warm air over land rises, creates a pressure difference, and pulls in cooler air from over a body of water (Lloyd, 2007). This creates a sea breeze and, during the day, the surface of the water will heat up. At night, the cooler air over land moves out over the water, creating a land breeze. Sometimes when the two meet, the water molecules will condense and form fog (Figure 5.13). Anyone who has visited San Francisco is familiar with this event.

Figure 5.13 Early morning fog over a pond.

Similar things can happen in mountains and valleys or hills and swales (Lloyd, 2007). The morning sun heats up the air over the flat valley floor, and the air rises up the slopes from midmorning to sunset (valley breeze). Then from midnight to sunrise, the heavier cool air lacks buoyancy and flows downward from the mountain slopes to the valley floor. It is not unusual for the moist air rising up the slopes to cause clouds to form, and depending on the amount of moisture, it could result in rain. Figure 5.14 and Figure 5.15 show the movement of odor-containing air moving up or down a slope. This should be considered when planning a search or watching how a dog works in a particular area at a specific time of day.

Figure 5.14 Updraft creating scent void (1) and remote scent cone (2) above the body. (From Rebmann, A., E. David, and M. H. Sorg. 2000. *The cadaver dog handbook: Forensic training and tactics for the recovery of human remains.* Boca Raton, FL: CRC Press. With permission.)

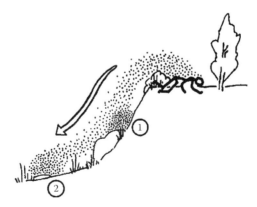

Figure 5.15 Downdraft creating remote scent pools below the body. (From Rebmann, A., E. David, and M. H. Sorg. 2000. *The cadaver dog handbook: Forensic training and tactics for the recovery of human remains.* Boca Raton, FL: CRC Press. With permission.)

Figure 5.16 Differences in the urban environment can affect odor dispersion.

Very locally, microweather conditions can occur. Air can rise on the sunny side of a tree or building (sometimes called the chimney effect) and once over in the shady side, can cool and then drop. Similar effects can occur in urban environments where similar microclimates may be created by heat radiating from buildings and pavement causing thermal uplift (Figure 5.16) (Lloyd, 2007). Something as simple as a tree line can wreak havoc with odor dispersion. The odor-containing air can blow toward the tree line (or other structure) at one level and then, due to an eddy or swirling effect, move out and away at a different level (Johnson, 1997). Odor can pass through openings or gaps in those structures to the opposite side, making localization or pinpointing very challenging for the HRD dog (Bulanda, 1994).

To recap. Warm air is less dense than cool air, air moves from high to low pressure, and that movement can create wind. Weather is influenced or made up of atmospheric pressure, moisture, and heat. HR chemicals are carried along with water vapor and can be transported by wind. Combinations of terrain, vegetation, location of the decedent, temperature, and wind are some of the factors that affect the availability of HR odor to the HRD dog.

Earth: The Dirt on Soil

Unless a chemical gets stuck on soil particles, think of soil as a sieve or filter (Figure 5.17). Sand, with its large particles with large empty spaces in

Figure 5.17 Soil types differ greatly in any given area.

between, allows material to pass through easily. Silty loams are basically 50% particulate and 50% porous space and can hold water more easily (Metting, 1993). Clay, with its small particles size and small empty spaces in between, holds particles in or they cannot pass through as fast or as easily. This probably applies to volatile organic compounds (VOCs) passing upward to the soil surface. Clay soil is likely to hold much of the gases from a buried body, as compared to sand. As already mentioned, the soil profile (especially layers) will change with human interference (Metting, 1993). It is important to remember that soil types can differ greatly in any given area, especially in wash or runoff areas (Stevenson and Cole, 1999).

Another way to describe soil is *hydraulic conductivity*. This explains how easy it is for water to pass through already water-saturated soil (Killpack and Buchholz, 2011). Hydraulic conductivity is the diffusion rate of saturated soil or Ksat. This tells how fast water will percolate or pass through soil in a given amount of time, usually in either in. or mm/sec. A high Ksat indicates a high rate of leaching or downward movement of substances from a body (such as purge fluid) on the surface into lower soil layers. Characteristics that affect the hydraulic conductivity includes the type of soil, the size of the pores or spaces between the pieces of dirt, texture, structures, and consistency, items covered in Chapter 3.

As we know, there are other things that hold or trap volatiles. Chemicals may be taken up into plants, soil "fauna" (including microbes and earthworms), or by organic matter in the soil (Stevenson and Cole, 1999).

Burials and Chemicals

A study done by Vass et al. (2004) analyzed chemical compounds that made it to the soil surface from buried decedents. The authors reported that it took 17 days for the first compounds to reach the soil surface from decedents buried 1.5 feet (0.45 m) deep. Many of the chemicals detected through the study appeared by one month after burial.

The depth of the grave had a significant effect on what classes of chemicals were produced, probably (as summarized by the authors) "due to the greater partial pressure of oxygen in shallow graves, which affects microflora" as well as the formation of chemicals. Barometric (air) pressure appeared to be a significant factor in the detection of various compounds. Table 5.3 shows the authors' interpretation of environmental effects on whether or not chemicals were detected.

Movement of chemicals down into the soil can be demonstrated by nitrogen. As a body breaks down, N is released both to the air (volatilizes) and downward to the soil as both organic N and inorganic as NH_4+ (ammonia). With the right conditions, the ammonium ion will bind to soil or continue to form NO_3- (nitrate). It is nitrate that can leach down through the layers of soil. Nitrification and mineralization (remember CO_2 and H_2O?) is affected by the pH, moisture level, level of aeration, and temperature of the soil. Nitrification increases as soil temperature increases.

Organic and inorganic chemical trace elements are absorbed and can be present in soils for a long time after the body is removed (Figure 5.18) (Larson, 2011).

Table 5.3 Possible environmental factors influencing where chemicals may be detected above or below a buried decedent[*]

| | Chemicals Detected above the Body Depends on: | | | |
Class	Barometric Pressure (BP)	Air Temperature	Rainfall	Optimum Surface Detection
Cyclic hydrocarbons	Yes	No	No	↑BP
Noncyclic hydrocarbons	No	Yes	No	Warm
Nitrogen compounds	?	?	?	—
Oxygen compounds	No	No	No	—
Acids/esters	Yes	No	No	BP
Sulfur compounds	No	No	No	—

[*] Adapted from Vass, A. A., et al. 2004. Decompositional odor analysis database. *Journal of Forensic Science* 49: 760–769. With permission.

Figure 5.18 Soil stained with purge fluid after the body was removed. (Photo taken by the author at the University of Tennessee's Forensic Anthropology Center.)

Ground Cover: Plants and More

As an HRD K9 handler, I have seen my dogs check out plants when working outside. Why are the dogs doing that?

So now its time to jump into the world of botany or the science of plants. When working outside, one will obviously encounter many different types of plants (Figure 5.19). Do they actually contribute or affect dispersion of odor from human remains? Let's look a little closer.

Plants grow by photosynthesis. This is the chemical process that plants use to combine water and CO_2 (requires energy) to make sugar and oxygen (Kohnke, 2011): six CO_2s plus six H_2O equals one sugar and six Os. Water serves as kind of the blood of the plant; by transporting nutrient molecules into the roots, up to the leaves, and then transporting waste molecules out.

Instead of respiration, plants "breathe" by transpiration. The definition of transpiration is "the process by which moisture is carried through plants from roots to small pores on the underside of leaves, where it changes to vapor and is released to the atmosphere" (USGS, 2011). Surfaces of leaves are covered with pores (stomata) that can open and close under the right conditions, being especially influenced by temperature.

To grow, the plant roots reach downward to pull water and nutrients from the soil (USGS, 2011). Water in the roots is pulled up to the shoots (leaves) by a decrease in hydrostatic pressure, caused by transpiration (State of Virginia, n.d.). As water leaves the leaves, the decrease in pressure allows water to circulate up to the shoots or leaves. When the leaves heat up, the water can change from a liquid to a vapor, the pores open, and gas is released. The cycle continues.

Figure 5.19 Different types of plants and different seasons can affect odor dispersion.

The amount of water released from the plant depends on the temperature, humidity, sunlight, wind, land slope, and water use. According to the USGS web site, transpiration can "actually contribute to the loss of moisture in the upper soil zone, which can have an effect on vegetation and food crop fields." The web site continues, "...an acre of corn gives off about 3000–4000 gal (11,400–15,100 l) of water each day, and a large oak tree can transpire 40,000 gal (151,000 l) per year" (Figure 5.20). On a warm, sunny day, a leaf can transpire many more times water than its own weight in one hour (USGS, 2011).

That explains water in plants, now what about other chemicals? Water will transport necessary minerals for the plant to the roots, including nitrogen, potassium, and phosphorus (Biology Encyclopedia, n.d.). It also includes movement of chemicals like fertilizers and herbicides (State of Virginia, n.d.). Soil active chemicals are pulled into the plant through the roots as they take up water and transpire. Environmental toxicologists study the potential effects of the uptake of chemicals into plants and the impact it may have on the plant itself or on animals that eat these plants (containing the chemicals). Researchers found that chemicals do accumulate in plants that were exposed through water and through biosolids (Wu et al., 2010; Tsiros, Ambrose, and Chronopoulou-Sereli, 1999).

Figure 5.20 Water cycling can include removal of water from soil by transpiration in plants and releasing it into the air.

One study of the uptake of putrescine into plants was done by Kakkar, Rai, and Nagar (1997). Putrescine was taken up into African violet petals, carrot cell cultures, and maize (corn) seedling roots. They found that uptake and movement of putrescine throughout plants was influenced by pH and concentrations of calcium, sodium, and potassium.

Therefore, it is not beyond the realm of possibility that (1) chemicals from a decomposing body are transferred to a plant via water uptake; (2) those chemicals are transported up into the plant's leaves; (3) some of those chemicals accumulate or are stored in the plant; and (4) when heated, some of those chemicals may be released in water vapor through transpiration. It also is possible that very dense low vegetation (of broad leaves) can block the upward movement of odor, making it less or unavailable to the dog (Bulanda, 1994).

After discussing water in plants, we now look at water in soil and beyond.

Water World

The Hydrologic Cycle

Hydrology describes the science of the movement, distribution, and quality of water throughout the Earth. We all learned the basic hydrologic cycle:

Figure 5.21 The water cycle: From lake to sky to lake.

water evaporates from a body of water, rises up, cools down, forms a cloud, the clouds rain, the rain ends on the water or land, and runs off the land to a lake or soaks into the soil, and ends up in a body of water (Figure 5.21). As already covered, evaporation from those bodies of water increases with higher temperatures and wind, and decreases when there is high humidity.

The Earth has all the water it will ever have. It cycles and recycles over and over and over again. About 97% of the Earth's water is contained in the oceans, only 3% exists as freshwater (Lloyd, 2007). Amazingly, only 0.3% of it is contained in lakes and rivers, which is remarkable when you think about it.

Hydrologic Measurements

In this section, we will look at a few definitions. The first is *vapor*. It is important to understand that water (H_2O) is the only molecule on Earth that can exist as a solid (ice), liquid (water), and gas (vapor). There are five major processes that water goes through: evaporation, transpiration, sublimation, condensation, and precipitation. You may wonder why this is important. It is important because many of the chemicals associated with HR follow water.

There are a few different ways that water is measured or described in weather terms. The first is *humidity,* which is simply how much water there is in the atmosphere. How much water the air can hold is related to both pressure and temperature. *Saturated vapor pressure* is the amount of pressure measured when the air is saturated with water (water and air in equal amounts). *Relative humidity* is a comparison of vapor pressure to saturated vapor pressure.

The *dew point* is the temperature at which the air becomes saturated with water and dew forms (by condensation). It is important to remember that when air saturated with water warms up (requiring energy), the water molecules warm up faster than the surrounding air, so they move faster. Then air with water vapor does not equal liquid water so more water can evaporate into the air. Saturation levels continually change.

Evaporation is the conversion of liquid water to gas water or vapor. As dry air comes into contact with a wet surface and energy is added, water evaporates. *Transpiration* is the process of plants taking water from the soil, up the roots to the leaves, and then to the air. With the help (and energy) of solar radiation and chlorophyll, water vapor diffuses out to the atmosphere. *Sublimation* is what happens when ice goes directly to vapor. With sublimation, there is no solid to liquid to vapor, just solid to vapor. *Condensation* requires the air to be "supersaturated" with water vapor. Then the water vapor molecules collide (because there are so many) and make larger molecules. The water molecules get even bigger when there are particles (like dust) to glob onto; this is how clouds or snow forms. *Precipitation* is what happens when those globs get bigger and bigger, and as they get heavier, they fall in the form of rain, sleet, or snow (Figure 5.22).

Figure 5.22 Precipitation: Heavy water molecules collect and fall back to the Earth.

Table 5.4 Physical characteristics of water*

Form	Drop Diameter (mm)	Density (Drops/ft²/sec)	Precipitation (in./h)
Fog	0.01	6264000	0.005
Drizzle	0.96	14	0.01
Light rain	1.24	26	0.04
Moderate rain	1.6	46	0.15
Heavy rain	2.05	46	0.60
Cloudburst	2.85	113	4.00

* Adapted from Herd, 2001.

Table 5.4 describes differences in various forms of water. Fog is kind of the clay of precipitation. As it has the smallest particle size and the densest of any type of precipitation, fog is really clouds on the ground.

There are two main types of fog: radiation and advection fog. Radiation or cooling fog forms when still, moist air cools and heat is pulled from the air to the Earth and the air cools to the point that dew forms. Radiation fog is seen in valleys as the cool air slides down the slopes. Advection fog is created when moist air flows over a cool surface (like water) and creates a fog. That fog can reflect sunlight and decrease localized warming (again San Francisco).

Other forms of H_2O include dew, frost, and snow. Dew forms after water condenses on the ground or objects. Frost is basically frozen dew, but it develops from sublimation. Snow can vary in how much water there is in it, ranging from dry (0%), to wet (3–8%), to slush (>15%). **The important point is that any condition that keeps warm moisture down, likely keeps odor down.**

Groundwater: Moving under Our Feet

There are three main types of groundwater flow. Shown in Figure 5.23, local flow can recharge or refill from "topographically high" areas, such as a pond

Figure 5.23 It is important to consider the movement of surface and groundwater because it can affect how an HRD K9 handler plans a search. (Diagram adapted from U.S. Geological Survey, www.USGS.gov)

on the side of a hill. If that pond is full, water can discharge out from the bottom to depressions that are lower, like a pond at a bottom of a hill. Waterways can fill either from surface runoff or from water traveling through soil layers and filling from the bottom.

In summary of impact of water: It is amazing how much the hydrologic cycle can affect how an HRD K9 handler plans a search. One should consider the cycle of water throughout the year in a specific location. In the midwestern United States, spring usually brings more rain, therefore, the waterways (rivers and lakes) recharge or fill from this precipitation. In the summer, those levels commonly drop because there is less rain and more evaporation (due to more direct sunlight) and more transpiration (from the increase in plant numbers and size). Movement of surface and groundwater with chemicals from decomposing human remains could carry odor a long way from the body. Figure 5.24 and Figure 5.25 demonstrate this.

People often ask how an HRD dog can locate a body under water. First scent sources likely include gasses (dissolved in water and in bubbles),

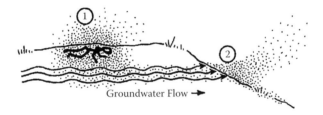

Figure 5.24 Primary and secondary scent pools/cones. (From Rebmann, A., E. David, and M. H. Sorg. 2000. *The cadaver dog handbook: Forensic training and tactics for the recovery of human remains.* Boca Raton, FL: CRC Press. With permission.)

Figure 5.25 Scent washes downhill with surface runoff to stream or secondary scent pool. (From Rebmann, A., E. David, and M. H. Sorg. 2000. *The cadaver dog handbook: Forensic training and tactics for the recovery of human remains.* Boca Raton, FL: CRC Press. With permission.)

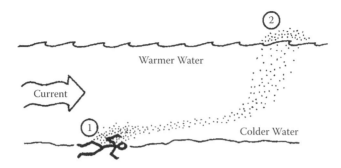

Figure 5.26 Movement of odor over water. (From Rebmann, A., E. David, and M. H. Sorg. 2000. *The cadaver dog handbook: Forensic training and tactics for the recovery of human remains*. Boca Raton, FL: CRC Press. With permission.)

liquids (plumes and droplets), and solids (tissues, etc.). Osterkamp (2011) described how this could happen. Potential scent transport processes for VOCs dissolved in water included molecular diffusion, turbulent diffusion, and entrainment in an upward flow of bubbles, buoyant liquid, and solids. He described studies of surface transport from seeps, kind of holes in the Earth's crust where oil and gas leak out and rise to the water's surface. As described, gas was released from the seeps as bubbles that were often coated with oil. Once the gas bubbles reached the water surface, they burst and left an "oily sheen" on the water. Oil droplets also floated up to the surface when released from the seeps, but at a much slower rate. Gas (often methane, in this case) was detected around 1 foot above the water's surface; this distance was increased by the action of waves and wind.

This is probably similar to what happens with odor from submerged decedents (Figure 5.26). In lakes without current, it is likely that the scent may pool in close proximity to the decedent. Windy conditions over lakes or rivers plus surface and subsurface currents can really affect odor dispersion. By using a "well-trained water search dog" to narrow the search area with other technology, recovery rates may be "greatly enhanced" (Judah, 2011).

Special Situations: Inside

When working inside, it is critical to be aware of the indoor environment and how ventilation systems, doors, and windows could potentially affect movement of odor (Mistafa, 1998). In unheated, cold buildings, odor could be concentrated around the source, just like outside. Hot temperatures can cause the odor to move up and out from the body, in a way, "filling up" and possibly saturating the room air.

Two cases were described by Preuss et al. (2006) in which bodies were located by "specially trained sniffer dogs" after being buried in concrete. One body was located approximately four months after the homicide; the other after eight months. In some situations, HR odor is able to disperse from bodies that are wrapped and buried in concrete.

What's New: Back to Orthogonal

Odor dispersion models have been developed and used for many years. Entomologists (bug guys) have studied pheromones of important pests in order to develop workable trapping methods and to distract or disrupt insect mating (Murlis, Elkinton, and Carde, 1992). They described "odor plumes" forming as the wind dispersed odor molecules from their sources. Their dispersion pattern was described as complex and "much like that seen in smoke plumes."

Computerized dispersion models also have been developed to predict where and how far odor plumes from Concentrated Animal Feeding Operations (CAFOs) can travel. Several models tested with "trained odor sniffers" have been used to determine where swine operations should be placed to minimize odors getting to neighboring homes (Guo et al., 2006). There are also many chemical dispersion models used by regulatory agencies to predict how chemicals can move through the environment as odors (Farrell et al., 2002) or if released from a "point source" (like a leaking underground storage tank). There is work underway to look at modifying or developing dispersion models to help the HRD dog handler plan how to most effectively deploy the HRD K9.

References

Bulanda, S. 1994. *Ready! The training of the search and rescue dog.* Portland, OR: Doral Publishing.

Ensminger, J. J., and I. E. Papet. 2012. Cadaver dogs. In *Police and military dogs: Criminal detection, forensic evidence, and judicial admissibility.* Boca Raton, FL: CRC Press, Chap. 19.

Farrell, J. A., J. Murlis, X. Long, L. Wei, and R. T. Carde. 2002. Filament-based atmospheric dispersion model to achieve short time-scale structure of odor plumes. *Earth and Environmental Science and Engineering* 2 (12): 143.

Guo, J., Y. Li, Q, Zhang, and X. Zhou. 2006. Comparison of setback models with field odor plume measurement by trained odor sniffers. Online at: http:engr.usask.ca/societies/csae/protectedpapers/c0536.pdf+scent+dispersion+modeling (accessed November 11, 2011).

Hammond, S. M. 2006. *Training the disaster search dog.* Wenatchee, WA: Dogwise Publishing.

Hazardous substances database. n.d. Bethesda, MD: National Institutes of Health's National Library of Medicine.

Herd, T. 2001. *Discover nature in the weather: Things to know and things to do.* Mechanicsburg, PA: Stackpole Books.

Johnson, G. R. 1997. *Tracking dog theory and methods.* Mechanicsburg, PA: Barkleigh Productions, Inc.

Judah, C. 2001. *Search and rescue dogs: Water search—finding drowned victims.* Brunswick, NC: Coastal Books.

Kakkar, R. K., V. K. Rai, and P. K. Nagar. 1997/1998. Polyamine uptake and translocation in plants. *Biological Plantarum* 40 (4): 481–491.

Killpack, S. C. and D. Buchholtz. 2011. Nitrogen cycle. Online at http://extension. missouri.edu/p/WQ252 (accessed October 10, 2011).

Kohnke, R. The role of water in the chemical processes of plants. Online at: http:// www.ehow.com/print/about_6558384_role-water-chemical-processes-plants. html (accessed on November 11, 2011).

Larson, D. O. A. A. Vass, and M. Wise. 2011. Advanced scientific methods and procedures in the forensic investigation of clandestine graves. *Journal of Contemporary Criminal Justice.* 27(2): 149–182.

Lloyd, J. 2007. *A pocket guide to weather: A complete guide to how the weather works.* Kent, UK: Parragon Publishing.

Metting, F. B. 1993. Structural and physiologic ecology of soil microbial communities. In *Soil microbial ecology: Applications in agricultural and environmental management.* New York: Marcel Dekker, Inc.

Mistafa, R. 1998. *K9 explosives detection: A manual for trainers.* Calgary, Alberta, Canada: Detselig Enterprises, Inc.

Murlis, J., J. S. Elkinton, and R. T. Carde. 1992. Odor plumes and how insects use them. *Annual Review of Entomology* 37: 505–532.

Osterkamp, T. 2011. K9 water searches: scent and scent transport considerations. *Journal of Forensic Science* 56 (4): 907–912.

Pearsall, M. D., and J. Verbruggen. 1982. *Scent: Training to track, search, and rescue.* Loveland, CO: Alpine Publications, Inc.

Preuss, J., M. Strehler, J. Dressler, M. Risse, S. Anders, S., and B. Madea. 2006. Dumping after homicide using setting in concrete and/or sealing with bricks— six case reports. *Forensic Science International* 159:55–60.

Purdue University. No date. Critical temperature and pressure. Online at: www.chem. purdue.edu/gchelp/liquids/critical.html

Rebmann, A., E. David, and M. H. Sorg. 2000. *The cadaver dog handbook: Forensic training and tactics for the recovery of human remains.* Boca Raton, FL: CRC Press.

Stevenson, F. J., and M. A. Cole. 1999. *Cycles of soil: Carbon, nitrogen, phosphorus, sulfur, and micronutrients.* New York: John Wiley & Sons.

Tsiros, I. X., R. B. Ambrose, and A. Chronopoulou-Sereli. 1999. Air-vegetation-soil partitioning of toxic chemicals in environmental simulation modeling. *Global News: The International Journal* 1(3): 177–184.

United States Geological Survey (USGS). The water cycle: transpiration. Online at: http://ga.water.usgs.gov/edu/watercycletranspiration.html (accessed on November 15, 2011).

Vass, A. A., R, R, Smith, C. V. Thompson, M. N. Burnett, D. A. Wolf, J. A. Synstelian, N. Dulgerian, and B. A. Eckenrode. 2004. Decompositional odor analysis database. *Journal of Forensic Science* 49: 760–769.

Virginia, State of. Herbicide fact sheet. Online at: http://www.dog.virginia.gov/mgt/herbidice-facts.html (accessed on November 15, 2011).

Weise, E., D, Vergano, and D. Rice. 2010. Icelandic volcano: The impact is broad, but could be worse! *USA Today*, April 20.

Wu, X., A. L. Spongberg, J. D. Witter, J. Fang, and K. P. Czajkowski. 2010. Uptake of pharmaceutical and personal care products by soybean plants from soils applied with biosolids, and irrigated with contaminated water. *Environmental Science and Technology* 44: 6157–6161.

Teaching Old Dogs New Tricks Using New Technology

6

Introduction

The goal of this chapter is to provide a broad background about some of the resources that can be used before or during a search to help the handler deploy the dog most effectively and efficiently. Although there are some advanced technologies that describe how to locate and recover bodies, they are beyond the limits of this book. So, after 10 years of handling human remains detection (HRD) dogs, it seemed appropriate to cover some of the basic information and sources I have found useful and, most importantly, easy to use.

Before we get into specifics, it is important to list some disclaimers. They include:

- This is not an inclusive list of resources to find background information useful in planning a search.
- Technologies referenced in this book are likely to be improved by the time of publication.
- Some of these technologies may apply to the United States only and not be available (or the same) as those available in other countries.

The best way to organize this chapter was to follow the format laid out in Chapter 5, outlining what resources may be useful to a handler before responding to a job. The first assumption will be the decedent is outside (Figure 6.1).

As covered in the previous chapter, the basic information needed by a handler before deployment is a plan that includes the search location and weather. The search location could be either on land or in the water. The span of weather from the time the decedent was reported missing to the date of deployment. Some of the resources needed before deployment include:

1. Location
   a. Maps
      i. Topography
      ii. Terrain (including cover, vegetation, etc.)
      iii. Hydrology (water flow)
      iv. Soil (type and layers)
2. Weather
   a. Past (temperature, precipitation, wind)
   b. Present (temperature, atmosphere [barometric] pressure, precip-
      itation, wind)
3. Examples of information and use in deployment tactics

Maps

Availability and detailed mapping is the area that probably has the most
information available and will continue to advance technologically. Because
of the pace of advances, this chapter will be limited to the basics, and limited
to maps in the United States. The basics of mapping for both land and water
searches will be covered.

Figure 6.1 Once a search area is provided, the work begins.

Definitions

The <u>Global Positioning System (GPS)</u> helps us determine our location. It was first developed in the mid-1970s by the U.S. Air Force and later became available to the general public. GPS depends on a number of satellites circling the Earth and "beaming radio signals back to the surface" (Letham, 2001).

A GPS receiver uses these radio signals to calculate location, which corresponds to a point on a map. GPS works in cloudy weather, but may not work under a thick canopy of trees or in cities where buildings can block the signals. It takes four satellites to line up and determine position; the accuracy of determining exact location is greater with better technology. Smaller and smaller receivers are being developed, small enough to be easily used in the field and on your dog.

<u>Geographic Information Systems (GIS)</u> is a form of technology that allows a user to combine "layers of digital data from different sources and to manipulate and analyze how the different layers relate to each other" (Eastern Geographic Science Center, 2011). What GIS does is allow people to look at maps in a completely new way and use that information for a variety of things. Some include soil use (we'll go into that later), emergency response planning (example: which rivers have flooded in the past depending on amount of rain), and "crime solving (police investigators link police record systems with geographic information to analyze crime patterns)" (EGSC, 2011).

<u>Latitude and Longitude (Lat/Long)</u> is the system that was developed long ago by sailors to navigate oceans. Latitude begins at the equator (0 degrees) and consists of parallel lines to the north and south (Hinch, 2007). These are not evenly spaced, but end at each of the poles with 90°N at the North Pole and 90°S at the South Pole.

Longitude are the imaginary lines that circle the globe around and through the poles. Longitude lines are parallel to each other and do vary in how far apart they are; e.g., at the equator, longitude lines are 69 miles apart but only 26 miles apart at the poles (Hinch, 2007). The 0 degree line (called the prime meridian) passes through Greenwich, England; lines from 0 to 180 degrees East or West of that line. By using the latitude and longitude, a precise spot can be determined.

Locations on these maps (as well as other GPS or GIS mapping systems) can be indicated by several different naming systems. Historically, most of us are familiar with the latitude–longitude system, but the <u>Universal Transverse Mercator (UTM)</u> and <u>Military Grid Reference System (MGRS)</u> also are used. What are they?

UTM is another locating method developed by the military that allows the round Earth map to be flattened and make each section the same. This grid "divides the entire world from 80 degrees south latitude to 84 degrees north latitude into 60 zones, each covering 6 degree of longitude" (Hinch, 2007). There are 60 zones, each 1,000,000 meters (~600 miles) wide, that are centered over the equator.

MGRS is a gridding system similar to the UTM.

General Maps

We are all familiar with a standard road map (Figure 6.2). But, as a handler, there must be more—and there is. One of the first things a handler may want to know is what does that area physically look like. So, off to the Internet for maps. One of the easiest to access and interpret is GoogleEarth. By going to the web site and typing in the address, one is immediately given an aerial view from satellite photographs (Google Map, 2011). It seems obvious, but it should be pointed out that this is *not* real-time photography, although in time, it may become that. With options of "map," "satellite," or "hybrid," one can easily get an idea of what type of area one may encounter.

With the ability to zoom in at street level in some locations on Google Map, it is possible to get an idea of buildings, trees, and streets, without leaving the comfort of your own chair. With the development of iPads, iPhones, android phones, etc., this part of our job is getting easier and easier.

From the satellite view, you can get a general impression of terrain, i.e., the general lay of the land. You can see if there are waterways, buildings,

Figure 6.2 The first step is determining location of the search area. (From Google Map, 2011.)

forests, etc.; the satellite view gives you a pretty good idea of what you may be in for (as well as mapping how to get there).

So, how do you do this? Steps for using the Internet to get surface and topographical imagery follow:

1. Internet address: maps.google.com
2. Selecting option "satellite" will produce maps showing roads, cover, etc. (Figure 6.3).
3. Selecting option "terrain" will produce geographical and topographical maps (Figure 6.4).
4. To "capture" the screen, the print screen function is the easiest way to obtain the images you want to use.

It is also a good idea to check web sites for the area that you are working in. Many (or most) cities and counties have GIS information and maps available. State web sites may have mapping information as well. In this day and age, finding maps should not be a problem.

Figure 6.3 Much information is available from basic Internet searches including satellite photos that provide a "lay of the land." (From Google Map, 2011.)

Figure 6.4 Terrain maps allow visualization of topography as well as what and how environmental factors could affect odor dispersion. (From Google Map, 2011.)

Topographical Maps

A topographical map is designed to show the variation in elevations that can be found in specific geographic locations through the use of contour lines. The contour lines show the shape and elevation; the closer together the lines are, the steeper the increase in elevation. Topographical maps are designed to show natural (hills, valleys, rivers, etc.) and manmade features (roads, buildings, etc.). The "amount of detail shown on a map is proportionate to the scale of a map" (EGSC, 2011).

To really know the lay of the land, a topography map should be consulted. Although some of the general mapping also gives elevation contours, some K9 handlers like to review more detailed mapping. Most maps on smart phone "apps" and private web sites are maps originally from the U.S. Geological Survey (www.usgs.gov). Although some of those maps may be outdated and do not show new manmade structures (buildings, roads, etc.), the basic contours, elevation, and physical details (waterways) are shown. It used to be necessary to purchase hard copies of these maps and carry them with you. With new technology, the USGS uses the tagline "science for a changing world" (USGS, 2011). This becomes obvious when looking at their web site where "a new generation of maps" describes "US Topo is the new generation of digital topographic maps from the USGS. Arranged in the traditional 7.5-minute quadrangle format, digital US Topo maps look and feel like the traditional paper" (USGS, 2011).

The USGS website is quite easy to use. They cover most (if not all) of the country and are based on the longitude–latitude system based on a scaling system (Figure 6.5). The USGS divided the country into precisely measured quadrangles, providing maps of the entire United States (Eastern Geographic Science Center, 2011). They are drawn to a specific scale: a 1:250,000 scale corresponds to 1 inch on the map equals 4 miles in the real world (Kjellstrom, 1994). A 1:62,500 scale is 1 inch on the map is about 1 mile, and 1:24,000 is about 1 inch to 2000 feet. The 1:24000 map is commonly called a "7.5-minute quadrangle map" because each map covers a four-sided area of 7.5 minutes of latitude and 7.5 minutes of longitude.

USGS maps are generally not very useful for urban searches because they usually are not detailed enough or updated as the "landscape" changes. This is where the Google (or similar, Mapquest) maps come to the forefront. Maps

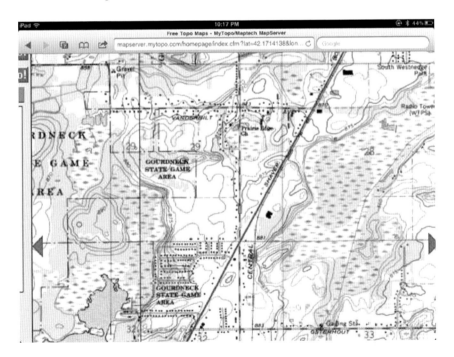

Figure 6.5 Topographical maps show differences in altitude and grade (how steep those differences may be). (From the U.S. Geological Survey.)

may be available from city or county web sites, from local fire departments, tax assessors, or county plat books (Young and Wehbring, 2007).

Because there may be some changes in terrain, it is sometimes helpful to look at older maps. One example I encountered was a request to search an area in a cold case. The actual site that came in on a tip indicated the decedent may have been placed in a lake. After looking up the location, I found that the lake evidently had turned into a marsh, which had filled in. Literally, we were working on what had been water.

Maps are available at no charge from the USGS Store and can be downloaded and stored on a computer. There are many features that allow the user to zoom in and out, select specific features of what one may want to or not want to see, and access "interactive capabilities with Google Map." There are even instructions on how to access and use the maps on the USGS web site.

For those with smartphones, there are several topographical applications available. Some are free, some cost, some work well, some don't. As quickly as things change, there are no recommendations for an outstanding topo app. <u>Just a quick note:</u> Remember that the more detailed the maps, longer download times and space requirements will be required. It is likely that the USGS or Google maps will provide sufficient detail for the K9 handler.

Hydrologic Maps

Water mapping is useful when planning a response to drownings in rivers or lakes (Figure 6.6). Google maps (or something similar) can provide a nice

Figure 6.6 Resources for finding out about what's under the water can include fishing guides, online resources, or local fishermen who know the waters. (Photo courtesy of FreeStockPhotos.biz.)

look over the body of water, helping to look at the basic lay of the land. For a more detailed look at lakes, topographical lake maps may be found in fishing guide books or again in smartphone apps that can provide some basic information. The maps provide information about water inlets and outlets, holes, and public access sites where boats can be launched.

Generally hydrologic maps are thought to include information about how water flows, either at the surface or below surface. Real-time mapping is available that provides information about flow rates and water height, all important information to understand possible implication on odor dispersion (Figure 6.7; DeShong: D5G Technology, 2011). Real-time information about coastal areas of oceans and the Great Lakes are available from the U.S. National Oceanographic and Atmospheric Administration's (NOAA's) web site (coastwatch.noaa.gov).

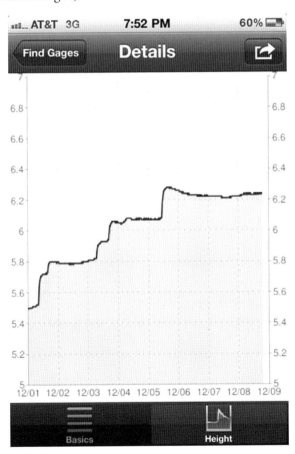

Figure 6.7 Some online applications allow real-time access to water flow and flood information (Flood Watch). (From Brian DeShong, D5G Technology. With permission.)

The USGS maintains a collection of hydrologic maps in the Hydrologic Investigations Atlases Series; these are maps that show a "wide range of water resource information, such as depth to groundwater, floods, irrigated acreage, producing aquifers, water availability, surface-water discharge, chemical or mineral content of water, and water temperature" (EGSC, 2011). The USGS web site has real-time mapping of streams, rivers, and lakes throughout the country. There are even smartphone apps for looking at real-time heights of rivers.

As of 2011, the USGS was in the process of developing groundwater atlases. As covered in the previous chapter, this type of mapping may be useful to understand dispersion of odor through surface or subsurface water pathways, especially in the case of buried decedents.

Soil Maps

As discussed in previous chapters, the role of soil on decomposition can be significant and it is extremely helpful to know what kind of soil is in the search area. In the United States, most of the soils have been mapped. One of the primary web sites that has soil maps is the Web Soil Survey (Figure 6.8). According

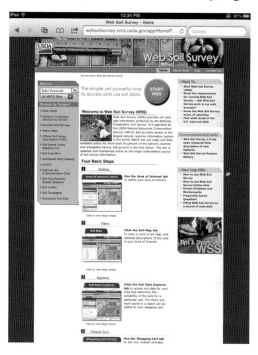

Figure 6.8 Soil information is available through online sources. This is the "front page" of a site that accesses soil maps from around the United States. (From the U.S. Department of Agriculture.)

to the web site, "The WSS provides soil data and information produced by the National Cooperative Soil Survey. It is operated by the U.S. Department of Agriculture (USDA) Natural Resources Conservation Service (NRCS) and provides access to the largest natural resource information system in the world. NRCS has soil maps and data available online for more than 95% of the nation's counties and anticipates having 100% in the near future. The site is updated and maintained online as the single authoritative source of soil survey information." That seems to say it all.

Google Earth also has some soil maps accessible at gelib.com/general-soil-map-of-the-united-states.html.

And, not to be outdone, there is a smartphone app called SoilWeb for iPhone (Figure 6.9). Although there is not yet a way to put in a location and it requires the user to be at the location where one wants to see the soil type, it has proved to be very useful (Beaudette and O'Geen, 2010). This app works using GPS to determine the location and will give a profile of the type and layers of soil that is literally underfoot (or within plus or minus a few feet) (Figure 6.10). SoilWeb also gives specific information about the soil and at what depth things like organic matter (%), percent of clay, percent of sand are

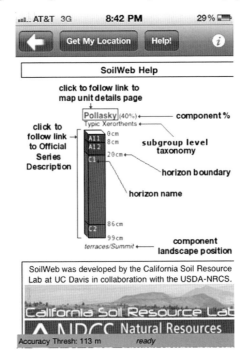

Figure 6.9 SoilWeb: A smartphone application that can be used in the field to find out the characteristics of what could be under one's feet. (From Beaudette, D. E., and A. T. O'Geen. 2010. *SoilWeb for iPhone*. University of California/Davis; Soil Resource Laboratory. With permission.)

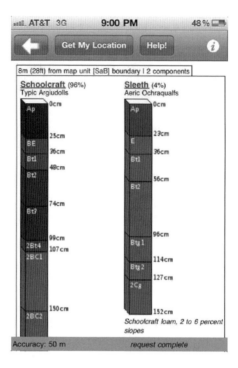

Figure 6.10 This shows a screen capture of the different types of soil at different depths (layers) at a specific site. (From Beaudette, D. E., and A. T. O'Geen. 2010. *SoilWeb for iPhone.* University of California/Davis; Soil Resource Laboratory. With permission.)

seen (Figure 6.11). Just knowing if you are looking for a decedent buried in clay versus sand can make a difference in odor dispersion, and, hence, may affect how you approach a deployment.

Weather or Not

Historical or Past

There are many web sites from which to gather past and present weather data. However, as in the maps section, I will cover only a few and those are in the United States. Like the maps section, there are also a number of apps that mainly provide current or forecasted information.

As covered in previous chapters, it is important to know the weather conditions that the decedent may have been in to determine what taphonomic changes could be expected. As previously covered, the contribution of temperature is the most critical factor affecting decomposition, therefore, this will be the primary focus. Weather may be less of a factor for affecting decomposition in old cases, but it certainly could affect odor dispersion.

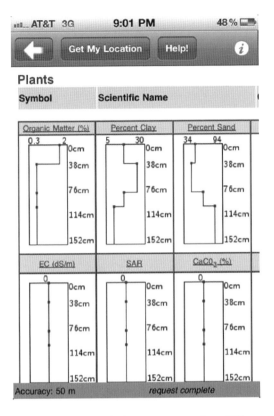

Figure 6.11 This is a capture of the concentration differences of different soil types. (From Beaudette, D. E., and A. T. O'Geen. 2010. *SoilWeb for iPhone*. University of California/Davis; Soil Resource Laboratory. With permission.)

The first step is finding the past daily temperatures to calculate accumulative degree days (ADD). There are several web sites, but one that is pretty informative is www.weatherunderground.com. It is a little tricky to find the historical data, but at this point in time, you have to click on the "Today's Almanac" tab, which is located next to the "Current Data" tab. Once you check on the Today's Almanac tab, you can put in the range of dates to get a calendar with the actual recorded high and low daily temperatures. At the very bottom in very small letters will be the View Detailed Hourly Forecast. After clicking on that, several graphs with temperature, rainfall, etc. will appear. Keep scrolling down and there is a place to put in a date. Although you put in a specific date (like the date the person was reported missing), weather data around that time pops up. There will be a tab for MONTH; selecting this will result in a monthly calendar with daily high and low temperatures (Figure 6.12). Although how you find the information may change, the daily high and low temperatures can be averaged to calculate ADDs. It

History : Weather Underground
www.wunderground.com/history/airport/KAZO/2007/3/12/MonthlyH

‹ Previous Month   ‹ 2006   **March 2007**   2008 ›   Next Month ›

Sunday	Monday	Tuesday	Wednesday	Thursday	Friday	Saturday
				1 Actual: 42|33, Precip: 0.29, Average: 40|21, Precip: 0.09	**2** Actual: 35|26, Precip: 0.02, Average: 40|22, Precip: 0.03	**3** Actual: 28|24, Precip: 0.00, Average: 42|24, Precip: 0.05
4 Actual: 30|24, Precip: 0.02, Average: 44|25, Precip: 0.13	**5** Actual: 30|10, Precip: 0.00, Average: 43|25, Precip: 0.05	**6** Actual: 18|6, Precip: 0.02, Average: 41|25, Precip: 0.01	**7** Actual: 28|15, Precip: 0.03, Average: 42|25, Precip: 0.08	**8** Actual: 32|8, Precip: 0.00, Average: 43|23, Precip: 0.05	**9** Actual: 46|19, Precip: 0.11, Average: 44|24, Precip: 0.05	**10** Actual: 46|33, Precip: 0.05, Average: 44|26, Precip: 0.08
11 Actual: 52|26, Precip: 0.00, Average: 46|25, Precip: 0.06	**12** Actual: 55|33, Precip: 0.00, Average: 47|26, Precip: 0.06	**13** Actual: 72|39, Precip: 0.00, Average: 46|27, Precip: 0.14	**14** Actual: 60|35, Precip: 0.15, Average: 46|29, Precip: 0.10	**15** Actual: 42|23, Precip: 0.08, Average: 48|29, Precip: 0.04	**16** Actual: 41|21, Precip: 0.00, Average: 47|27, Precip: 0.11	**17** Actual: 39|19, Precip: 0.00, Average: 47|25, Precip: 0.07
18 Actual: 44|19, Precip: 0.00, Average: 48|28, Precip: 0.06	**19** Actual: 54|33, Precip: 0.04, Average: 49|31, Precip: 0.05	**20** Actual: 46|27, Precip: 0.00, Average: 48|29, Precip: 0.07	**21** Actual: 68|26, Precip: 0.37, Average: 48|28, Precip: 0.13	**22** Actual: 64|41, Precip: 0.22, Average: 48|27, Precip: 0.10	**23** Actual: 53|29, Precip: 0.00, Average: 49|29, Precip: 0.03	**24** Actual: 64|44, Precip: 0.19, Average: 47|27, Precip: 0.06
25 Actual: 76|50, Precip: 0.00, Average: 49|27, Precip: 0.06	**26** Actual: 79|64, Precip: 0.00, Average: 49|30, Precip: 0.06	**27** Actual: 78|60, Precip: 0.14, Average: 50|29, Precip: 0.10	**28** Actual: 57|48, Precip: 0.01, Average: 52|32, Precip: 0.18	**29** Actual: 62|44, Precip: 0.00, Average: 52|34, Precip: 0.11	**30** Actual: 70|39, Precip: 0.00, Average: 51|54, Precip: 0.12	**31** Actual: 51|44, Precip: 0.04, Average: 54|32, Precip: 0.05

Figure 6.12 To help calculate accumulated degree days (ADD), historical weather (including average daily temperature) can be used. (Image and data from weatherunderground.com. With permission.)

is important to remember to convert the Fahrenheit to Celsius if you are comparing it to information from the published literature.

An easier site to use is sponsored by the NOAA. Like the USGS, this web site has a treasure trove of information. The National Climatic Data Center allows one to search for weather data for a specific location by using the following instructions:

1. Go to http://www.ncdc.noaa.gov/cod-web/search.
2. Use either map search or data search.
3. Map search allows for searching via map of location (Figure 6.13) and gives fast, simple information including temperatures, precipitation, etc.
4. Data search is useful for a more detailed information search in a specific area (Figure 6.14).

Another online site for historical data is www.weatherpages.com/wxhistory.html. This site allows a user to get data from the weatherunderground.com web site. It also allows one to get hour-by-hour weather data, but you have to indicate a specific weather station of interest.

It also is useful to look at trends in precipitation: Was there any significant rain events that could drive odor down into subsurface flow or runoff? When did these occur? Did it occur just before your deployment? It may be

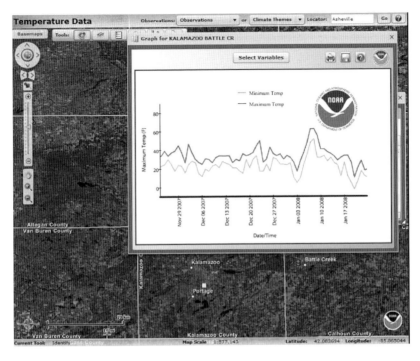

Figure 6.13 Information about historical temperatures can be done by searching by location. (From the National Oceanic and Atmospheric Administration (NOAA).)

Figure 6.14 NOAA allows data searching for more detailed information. (From the National Oceanic and Atmospheric Administration (NOAA).)

helpful to be aware of these events when planning how you may deploy the detector dog.

Current/Predicted Weather

This is the easiest information to find. Thanks to the Weather Channel, weather web sites, and smartphone apps, current and predicted conditions are available at the touch of a button. What information you want to look for is, of course, temperature, wind (direction and speed), barometric pressure, and precipitation. HR handlers often have the option of delaying deployment until conditions are favorable. Favorable conditions are cool but warming (not cold, not hot), moist conditions with light winds.

At the least, being familiar with past and present weather would be helpful to be able to explain what taphonic changes the decedent may have and how the dog may work in these conditions. It is very important for the handler to know this to help interpret how the dog is working and if there is anyway the handler can help the dog.

Behavior of the Missing

When looking for a missing person (usually referring to a person who gets lost or takes off), it is helpful to understand some of the commonly shown behaviors of that individual or "group" of people. The point is that people get lost in different ways. For example, very young children may be distracted and follow butterflies, may not respond to loud male voices calling their names, etc. Alzheimer's patients get lost easier than the normal person. Does the person have a history of getting lost or taking off? Does anyone know where they went? Are they physically and mentally prepared to deal with different environment conditions?

This information is extremely useful in planning a deployment. There are some valuable sources of information that can be used that describe some of the basic behaviors that certain "types" of people may follow when they are lost. An excellent one is *Analysis of Lost Person Behavior: An Aid to Search Planning* (Syrotuck, 2000). Endorsed by the National Association for Search and Rescue (NASAR), this book provides very helpful information. Another is the web site http://www.dbs-sar.com/LPB/lpb.htm and the book *Lost Person Behavior* (Koester, 2008). Other organizational web sites (e.g., Alzheimer's Foundation of America) often contain useful background information.

Using This Information

Using this information is the next step. The following is from a case that we were involved with several times over a four-year period. This case is presented in Chapter 7 as Case 5.

The History
Location: Central Midwest United States, in a rural area where an individual (adult, middle-aged male) with a known substance abuse issue threatened suicide. He left about 11 p.m. the night before and was thought to have walked out behind the residence into a wooded, swampy area. A tracking dog was called that was able to determine a direction of travel.

Our services were requested four days later. Maps were provided, and the area was just as described—the high point was at the back of the residence, with a pretty good slope down to a very large wooded swamp that led into a fairly wide river. The weather at the time the person disappeared was cool and overcast. There had been above-average rainfall during the preceding month, so the ground was saturated. The weather at the time of deployment was cool, overcast, with significant ground fog and no wind. I advised the detectives that the probability of detection was likely limited because the environmental conditions were certainly not favorable for detection, especially from a distance in a very large search area. That being said, we worked. As possible, we worked the area with a fairly tight grid due to weather conditions. At the end of the search, the dog showed no indications. I suggested that we return when conditions improved or to bring in other certified operational HRD K9s if information showed a continued search in the area was indicated.

Time passed; conditions improved; using mapping and weather data, we planned and we searched. More "tips" came in. We searched. More time passed. This same scenario continued off and on over the next few years with no success in locating the missing person. Then about four and a half years later, a surveyor working along the river hit something hard with a probe. Assuming it was a rock or stick, he pulled up a human femur. The missing had been found. The decedent had evidently traveled farther than expected, must have gotten stuck in some very gooey mud, and was stopped. Most of the skeleton was recovered along with some personal effects and the decedent was identified. As can be seen in Chapter 7, the skeletal remains picked up the dark pigments of the black, sticky mud in which they were found. The bones were somewhat displaced, likely due to the freezing/thawing cycle that took place over several years.

Take home story: As with many cases, we were never physically close to where the decedent was found (closest was about one and a half to two miles away) and, even if we had been close, the weather on the first day could have negatively impacted (inhibited) any odor dispersion. The low-lying area, the decedent in water, probably covered in thick mud, on a foggy morning equals a truly challenging deployment.

Figure 6.15 Whatever the location, it is helpful to get as much information before deploying, but plans may change depending on what is happening at the time of deployment.

Summary

Although technology has and continues to develop, many times the success of a search comes down to being in the right place at the right time and under the right conditions. Planning and knowing what to expect should help the K9 handler effectively deploy the HRD dog (Figure 6.15).

References

Beaudette, D. E. and A. T. O'Geen. 2010. *SoilWeb for iPhone*. Soil Resource Laboratory, University of California/Davis. Online at: http://casoilresource.lawr.ucdavis.edu (accessed October 15, 2011).

DeShong, B. 2011. *FloodWatch: A smartphone application*. D5G Technology.

Eastern Geographic Science Center (EGSC)/ U.S. Geological Survey (USGS). Topographic mapping. Online at: http://egsc.usgs.gov/sib/pubs/boodlets/topo/topo.html (accessed October 20, 2011).

Google Earth Library (GELIB). Terrain maps (TOPOMATT, May 22, 2010). http://www.gelid.com/terrain-maps.html (accessed on October 24, 2011).

Hinch, J. *Outdoor navigation with GPS*, 2nd ed. 2001. Berkeley, CA: Wilderness Press.

Kjellstrom, B. 1994. *Be expert with map & compass: The complete orienteering handbook.* New York: Macmillan General Reference/Simon & Schuster Macmillan Company.

Koester, R. J. 2008. *Lost person behavior: A search and rescue guide on where to look—for land, air and water.* Charlottesville, VA: dbS Productions, Inc.

Letham, L. 2001. *GPS made easy: Using global positioning systems in the outdoors*, 3rd ed. Seattle, WA: The Mountaineers.

NOAA (National Oceanic and Atmospheric Administration)/Department of Commerce. 2011. Courtesy of the National Oceanic and Atmospheric Administration Central Library Photo Collection. Online at: http://www.lib.duke.edu/libguide/cite/web_pages.htmNavionics (accessed October 24, 2011).

Syrotuck, W. G. 2000. *Analysis of lost person behavior: An aid to search planning.* Mechanicsburg, PA: Barkleigh Publishing.

United States Geological Survey (USGS). *A new generation of maps.* Online at: http://nationalmap.gov/ustopo/index.html (accessed on October 24, 2011).

Young, C. S., and J. Wehbring. 2007. *Urban search: Managing missing person searches in the urban environment.* Charlottesville, VA: dbS Productions, Inc.

A Picture Is Worth a Thousand Words

7

Introduction

This chapter was only possible because of the generosity of several departments and their recognition of the importance of education for those involved in death investigation.

Note: It is important to point out that these photos are sensitive in nature and should *not* be viewed by nonprofessionals. The loss of life is tragic, sometimes horrific, and is never taken lightly. The opportunity to use these photos for educational purposes is greatly appreciated.

Six cases are covered that include decedents exposed to different environmental conditions in the northern midwestern United States. The cases are varied and took place in the winter or summer, were of long-term or short-term duration, and happened on land or water.

Case 1: Cold Weather/Short Term/Land

History

Adult female disappeared mid-January, 2007. There was a history of substance abuse at the time the decedent went missing and the place she was last seen was at a retail store where closed-circuit television showed the decedent drinking. She had a history of walking from her home to the store, a trip of approximately 1.5 miles each way.

Area

Figure 7.1 shows an aerial photograph where the decedent was found. The area was wooded with a rough walking trail passing through it.

Figure 7.1 (See color insert) Aerial photo of area where Case 1 decedent was located. (From Google, 2011.)

Weather

Weather initially mild for winter in the first few days, followed by dropping temperatures and significant snow fall. This continued until the mid-March thaw. Accumulated degree days (calculated from January 15–March 15): 44°C.

Search Efforts

The area was initially searched without success. A human remains detection (HRD) K9 was not available until six days later, at which time more than one foot of snow had accumulated on the ground. Tactical planning sessions were conducted to determine which areas to search when the weather cleared.

Results

The area the body was found was not searched at the time the decedent went missing because search efforts were concentrated closer to the point she was last seen. Deep snow cover persisted until mid-March when rains helped with a quick melt. Before an HRD K9 was deployed, an officer searching the area on foot located the decedent approximately 25–30 feet from a path (Figure 7.2). An article of clothing (scarf) and a bottle are present in the foreground. The decedent (dressed in dark winter clothing) is visible in the upper

Figure 7.2 (See color insert) Decedent located about 25 feet to left of path.

Figure 7.3 (See color insert) Closer view of decedent dressed in dark clothing.

left corner of the photo. Figure 7.3 shows a closer view of the decedent as found with her face down and head against a small tree. Once the decedent was rolled up, an "outline" of fluid was found under the body. This was not purge fluid (due to lack of decomposition), but was likely condensation that was created when the body came into contact with the ground (Figure 7.4). Figure 7.5 shows the minimal decompositional changes, limited only to the

Figure 7.4 (See color insert) Fluid on ground likely due to condensation of fluid under decedent.

Figure 7.5 (See color insert) Minimal decomposition limited to the face, which was in contact with soil.

face. Because the weather was cold and the decedent was fully covered, only the area that had contact with the ground showed any decomposition. It is likely that the decomposition resulted from soil microbes that were probably still somewhat active in the not yet frozen soil.

Case 2: Hot Weather/Short Term/Land

History

Adult male reported as homeless was last seen in mid-April 2008.

Area

Figure 7.6 shows an aerial photo where the decedent was found. The area was wooded and was just behind a number of small businesses.

Weather

It was generally hot and dry from mid-summer to mid-October. There were a few exceptional rainfalls: Total rainfall in July was 3.24 inches; August, 1.29 inches; September, 11.12 inches (10.5 inches in a 36-hour period); and 1.1 inches through mid-October. Accumulated degree days (calculated from mid-July to mid-October): 3419°C.

Figure 7.6 (See color insert) Aerial photo of area where Case 2 decedent was located. (From Google, 2011.)

Search Efforts

The decedent was located in mid-October by a passer-by who was reportedly looking for areas in which to hunt.

Results

Figure 7.7 and Figure 7.8 show the body next to a tree. The decedent committed suicide by ligature strangulation. There was a belt found around the neck with the decedent in a sitting position. Once the belt was cut, the body fell forward resulting in the head oriented in a slightly upside down position with the chin facing up. The eye and nose openings are visible. The head was skeletonized likely due to a high level of flies and insect activity. Some of the skin (on top of the skull) was intact, but leathery in appearance (Figure 7.9). Most of the body was covered by clothing and, thus, was not in contact with the ground or to free accessibility by insects. The hands were partially mummified.

Figure 7.10 shows the imprint on soil created by decompositional fluid. Figure 7.11 shows a closer view along with lots and lots of pupal casings from numerous flies.

Insects collected at the scene and submitted to a forensic entomologist provided information that the decedent had likely died in mid-July. Environmental conditions were similar to those shown in Chapter 3 with Horace, the pig.

Figure 7.7 (See color insert) Decedent sitting at base of tree.

Figure 7.8 (See color insert) Closer view of decedent who apparently committed suicide by ligature.

Figure 7.9 (See color insert) After ligature was released, the decedent fell forward, showing partial mummification of skin over head along with partial skeletonization.

Figure 7.10 (See color insert) Imprint of soil likely due to decompositional fluid.

Figure 7.11 (See color insert) A closer view of fluid in soil along with many dark fly pupal casings.

Case 3: Hot Weather/Short Term/Water

History

Young adult male last seen on June 10 at a county park with a lake.

Area

The area the decedent was found in was a county park with a lake that had a swimming beach (Figure 7.12).

Weather

The decedent was reported missing on June 10 and found on June 13. The daytime highs the first two days were in the high 80s (F) and in the high 60s (F) the last two days. The county health department had tested the water during the time the decedent was missing; water temperature was reported at 79°F. The accumulated degree days (ADD) (calculated from June 10 to 13): 141°C.

Search Efforts/Results

The decedent was found mid-afternoon on June 13 (Figure 7.13). He was "floating in a head down position with his legs hooking the handicap handrail

Figure 7.12 (See color insert) Aerial photo of area where Case 3 decedent was located. (From Google, 2011.)

that is approximately 2 feet down from the surface of the water" (Figure 7.14). The water depth where he is found was about 5 feet deep.

The decedent was in a state of rigor (Figure 7.15) and showed early signs of gas bullae in the skin along with skin slippage (Figure 7.16). There was some degree of bloating that likely allowed the decedent to resurface.

Figure 7.13 (See color insert) Decedent visible floating in lake off public beach.

Figure 7.14 (See color insert) Closer view of decedent.

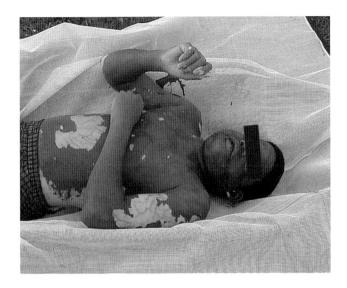

Figure 7.15 (See color insert) Decedent in persistent rigor.

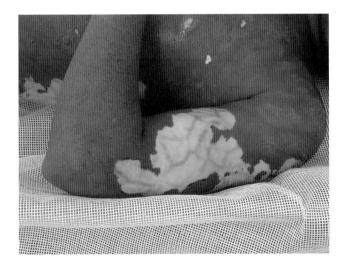

Figure 7.16 (See color insert) Gas bullae with skin slippage.

Case 4: Long Term/Land

History

An adult female was last seen in mid-July after prematurely leaving an emergency room following being admitted for possible drug overdose.

Figure 7.17 (See color insert) Aerial photo of area where Case 4 decedent was located. (From Google, 2011.)

Area

The area where the decedent was found was wooded, but in very close proximity to a housing complex (Figure 7.17).

Weather

Not evaluated as two and a half years passed between time last seen and located. Accumulated degree days were not calculated.

Search Efforts

The history of search efforts was not available. The decedent was located in January, almost two and a half years later by a dog owned by a resident of the nearby housing complex.

Results

Figure 7.18 shows a view of the area where the decedent was located (area near the fence). In a closer view, some bones can be seen along the fence (Figure 7.19) where they have been flagged (Figure 7.20). As would be

Figure 7.18 (See color insert) Skeletal remains located near center of photo.

Figure 7.19 (See color insert) Closer view of skeletal remains.

expected, the body was completely skeletonized and bones were scattered in the general area. Some of the bones were very white in appearance, while others took on the color of the soil (Figure 7.21).

Once the obvious bones were removed, an HRD K9 helped locate more bones in the area (Figure 7.22).

Figure 7.20 (See color insert) Flags mark bones.

Figure 7.21 (See color insert) Some bones bleached white, while others took on color of soil and surrounding vegetation.

Figure 7.22 (See color insert) Some of the other bones located with help of HRD K9.

Case 5: Long Term/Land–Water

History

An adult male threatened suicide by possible drug overdose after an argument late one night in late May 2006. The decedent knew the area well and was known to walk in the woods.

Area

Figure 7.23 shows an aerial view of the area. The house backed up to a wooded, swampy area that was separated by an elevated two-track utility right-of-way that ran in generally a north–south direction. The two-track was the high point of the area that sloped eastward down to a large swamp and wetlands area. It eventually ran into a river.

Weather

Not evaluated because four and a half years passed between the time the decedent was last seen and the location. ADDs were not calculated.

Search Efforts

A tracking dog was able to establish the direction of travel (north along the two-track) a few hours after the man left the residence. The K9 lost the

Figure 7.23 (See color insert) Aerial photo of area where Case 5 decedent was located. (From Google, 2011.)

track after about a half mile. Four days later, an HRD K9 was brought in and searched the area; conditions were not favorable for odor dispersion as there was a heavy ground fog. About a week later, two HRD dogs were deployed, covering a larger area without success. Intermittent searches were made throughout the next several years without finding the decedent.

The decedent was located in September 2010, about four and a half years after disappearing, by a surveyor, several miles from the areas that were previously searched. The decedent was located when the worker was probing deep into the mud of an overflow flood area (Figure 7.24).

Results

Figure 7.25 shows the area the decedent was located in (area near the "puddle"). In a closer view, some bones (ribs) can be seen that have taken the color from the surrounding mud (Figure 7.26). The body was completely skeletonized; clothing also was recovered from the mud in that same area (Figure 7.27).

This is one of those cases that shows how large the world is and how comparatively small a human is. His body would have gone undiscovered had it not been for a surveyor slogging through the mud.

Figure 7.24 (See color insert) Wooded swampy area where decedent was reported missing.

Figure 7.25 (See color insert) Center of photo: Decedent's remains recovered in area under mud.

Figure 7.26 (See color insert) Ribs from the decedent located in the mud.

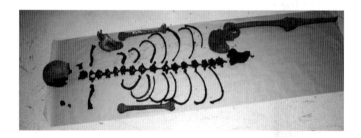

Figure 7.27 (See color insert) Remains recovered.

Case 6: Long Term/Burial

Although all previous cases were tragic, the last one (to me) is the saddest because of the lack of value that a mother placed upon her own child. The way this child was treated before and after death was almost beyond belief.

History

This case concerns a mother of seven with a history of domestic abuse targeted at her children, especially a four-year-old female. According to information from a family member, the young girl died sometime in late November from injuries inflicted by her mother. If that wasn't bad enough, she stashed the girl's body under the kitchen sink for a couple of days. She then put the girl's body in the trunk of her car where the body remained until May of the following year. The mother had an accident (with other children in the car), resulting in the car being totaled. It was towed to an impound car lot where it was parked. It is necessary to note that, in photos of the car, it was so full of garbage and junk that it would have been easy to miss the body of a small child stashed in the trunk. The mother returned about a week later, claiming to need to get a few things from the vehicle, and she did.

Area

Figure 7.28 shows an aerial view of the area where the mother drove out and buried her daughter in a thickly wooded area in late May.

Weather

In June of the following year, a family member informed officials of what had happened to the young girl. Because a total of one and a half years had passed (around six months in the car trunk over the winter and about one year buried), ADDs were not calculated.

Figure 7.28 (See color insert) Aerial photo of area where Case 6 decedent was located. (From Google, 2011.)

Search Efforts

A family member took authorities to the general area where the young girl had been buried. Extensive ground searching was done with law enforcement (LE) officers without success (Figure 7.29). Later that day, an HRD K9 was deployed and was able to locate the clandestine grave.

Recovery efforts revealed the top of a garbage bag about 12-inches deep (Figure 7.30). The decedent was wrapped in three plastic bags and dumped into a grave that was about three-feet deep. Upon opening the bag, the skull of the decedent was visible (Figure 7.31). Once removed, the body showed some degree of mummification and adipocere formation (Figure 7.32). The body likely did not go through the putrefactive process as it was closed in a trunk and went through several freeze–thaw cycles between November and May. Then, when placed in the plastic bags and buried, the moist conditions with low oxygen levels resulted in adipocere formation.

Summary

Loss of life is a tragedy. It is hoped that these cases provide the reader with some useful examples demonstrating the environmental factors that influence the taphonomic process.

Figure 7.29 (See color insert) Decedent was buried in this wooded area off the road.

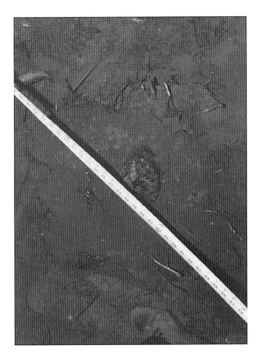

Figure 7.30 (See color insert) Layers of soil removed, revealing top of plastic bag.

Figure 7.31 (See color insert) Top of decedent's head showing hair and adipocere.

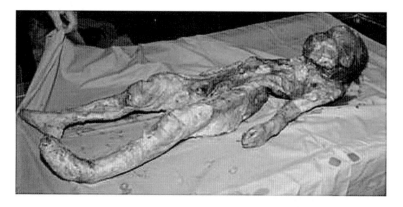

Figure 7.32 (See color insert) Some degree of mummification and adipocere seen in decedent.

Beginning of the End 8

So, there has been a lot of information thrown at the reader in this book. This last part will capture the highlights of each chapter and hopefully pull it together. Because this book is a guide, it covered only the highlights of a number of different scientific areas.

Chapter 1: The locating tool. Basic anatomy and physiology of canine olfaction were covered along with some of the unique characteristics that help a dog do what they do so well. The objective was to give the K9 handler an understanding of how their dog works as a locating tool.

Chapter 2: Forensic taphonomy. Normal (living) cell structure was covered along with the requirements a cell needs to live. As those resources become depleted, the processes that a dying cell goes through were reviewed. Expanding from the cellular level to the systemic and, finally, total body, the stages of decomposition were reviewed. Highlights of the mortices (algor, livor, and rigor) were covered.

Chapter 3: Environmental variables affecting decomposition. This chapter dove a little deeper in the taphonomic factors that can alter how the process takes place and what the handler and/or investigator may see. By touching upon effects of weather and location as well as the contributions of arthropods ("big bugs"), scavengers, and predators, the reader should understand why and what taphonomic changes were seen. The contributions of Hank and Horace (the pigs) were presented, providing a pictorial "display" of the process and how the environment affects the physical outcome.

Chapter 4: The chemistry of odor. This chapter outlined the "microenvironmental" factors that further affect decomposition. By understanding the contributions of "little bugs" (microbes from the body and/or environment), the reader should understand the products of decomposition and have a general background in the organic chemicals that can/may make up the odor profile of human remains (HR). This background should help the handler explain the complexity of the work that human remains detection (HRD) dogs do and the almost limitless profile that the dogs may be expected to detect.

Chapter 5: Environmental variables affecting the odor of decomposition. This chapter considered some of the variables that can affect dispersion of odor from a decedent in different environments. By comparing diagrams from *The Cadaver Dog Handbook* (Rebmann, David, and Sorg, 2000), the reader should understand how weather, topography, ground cover, and terrain can affect odor dispersion, helping to deploy the dog most effectively.

Chapter 6: Tactics and planning—using technology for deployment. This chapter covered some of the newer technologies that can be used before or while deploying the HRD dog. The focus is on different types of maps and weather (past and present).

Chapter 7: A picture is worth a thousand words. This is the chapter that tied all the others together by providing case reports about decedents who have been found in different locations in a variety of environmental conditions. After learning how these variables can affect how and what is found, the handler and/or investigator should be better prepared to do his/her job.

REFERENCE

Rebmann, A., E. David, and M. H. Sorg. 2000. The cadaver dog handbook: Forensic training and tactics for the recovery of human remains. Boca Raton, FL: CRC Press.

Index

Figure 7.1 Aerial photo of area where Case 1 decedent was located. (From Google, 2011.)

Figure 7.2 Decedent located about 25 feet to left of path.

Figure 7.3 Closer view of decedent dressed in dark clothing.

Figure 7.4 Fluid on ground likely due to condensation of fluid under decedent.

Figure 7.5 Minimal decomposition limited to the face, which was in contact with soil.

Figure 7.6 Aerial photo of area where Case 2 decedent was located. (From Google, 2011.)

Figure 7.7 Decedent sitting at base of tree.

Figure 7.8 Closer view of decedent who apparently committed suicide by ligature.

Figure 7.9 After ligature was released, the decedent fell forward, showing partial mummification of skin over head along with partial skeletonization.

Figure 7.10 Imprint of soil likely due to decompositional fluid.

Figure 7.11 A closer view of fluid in soil along with many dark fly pupal casings.

Figure 7.12 Aerial photo of area where Case 3 decedent was located. (From Google, 2011.)

Figure 7.13 Decedent visible floating in lake off public beach.

Figure 7.14 Closer view of decedent.

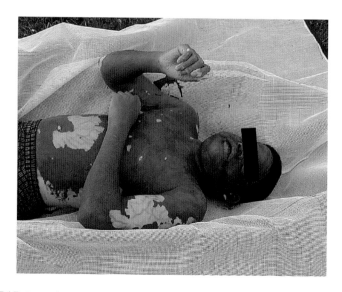

Figure 7.15 Decedent in persistent rigor.

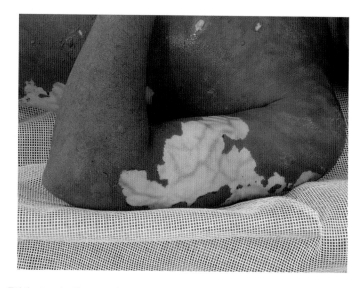

Figure 7.16 Gas bullae with skin slippage.

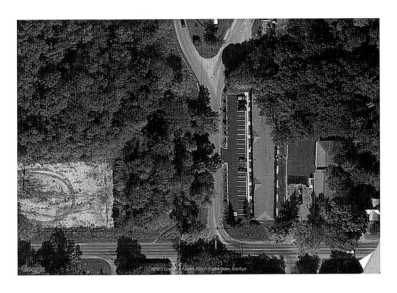

Figure 7.17 Aerial photo of area where Case 4 decedent was located. (From Google, 2011.)

Figure 7.18 Skeletal remains located near center of photo.

Figure 7.19 Closer view of skeletal remains.

Figure 7.20 Flags mark bones.

Figure 7.21 Some bones bleached white, while others took on color of soil and surrounding vegetation.

Figure 7.22 Some of the other bones located with help of HRD K9.

Figure 7.23 Aerial photo of area where Case 5 decedent was located. (From Google, 2011.)

Figure 7.24 Wooded swampy area where decedent was reported missing.

Figure 7.25 Center of photo: Decedent's remains recovered in area under mud.

Figure 7.26 Ribs from the decedent located in the mud.

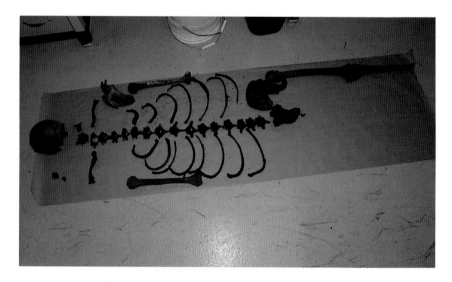

Figure 7.27 Remains recovered.

Figure 7.28 Aerial photo of area where Case 6 decedent was located. (From Google, 2011.)

Figure 7.29 Decedent was buried in this wooded area off the road.

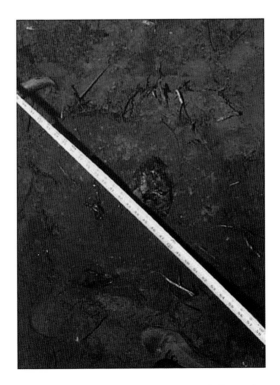

Figure 7.30 Layers of soil removed, revealing top of plastic bag.

Figure 7.31 Top of decedent's head showing hair and adipocere.

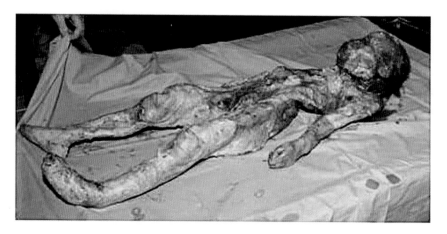

Figure 7.32 Mummification and adipocere seen in decedent.